ENVIRONMENT AND TRANSPORT IN ECONOMIC MODELLING

FONDAZIONE ENI ENRICO MATTEI (FEEM) SERIES ON ECONOMICS, ENERGY AND ENVIRONMENT

This series serves as an outlet for the main results of FEEM's research programmes in the areas of economics, energy and environment.

The titles published in this series are listed at the end of this volume.

Environment and Transport in Economic Modelling

Edited by

ROBERTO ROSON

Department of Economics, Ca' Foscari University, Venice, Italy

and

KENNETH A. SMALL

Department of Economics, University of California, Irvine, CA, USA

KLUWER ACADEMIC PUBLISHERS

DORDRECHT / BOSTON / LONDON

Library of Congress Cataloging-in-Publication Data

A C.I.P. Catalogue record for this book is available from the Library of Congress.

ISBN 0-7923-4913-X

Published by Kluwer Academic Publishers,
P.O. Box 17, 3300 AH Dordrecht, The Netherlands

Sold and distributed in North, Central and South America
by Kluwer Academic Publishers,
101 Philip Drive, Norwell, MA 02061, U.S.A.

In all other countries, sold and distributed
by Kluwer Academic Publishers Group,
P.O. Box 322, 3300 AH Dordrecht, The Netherlands.

Printed on acid-free paper

Printed in Great Britain

Table of Contents

CHAPTER 1

Introduction: Modelling Environment and Transport

Roberto Roson and Kenneth A. Small

Recent years have seen an outpouring of literature on the relationships between transport and environmental policy. Besides a healthy number of contributions through the normal channels of journals, reports and monographs, entire new journals have been created and several edited collections of articles have appeared. This is no wonder, given the perceived importance of environmental problems in modern developed economies and the role the transport sector plays. In Europe, the trend is further strengthened by the importance of harmonizing both environmental and transport policies as part of unification, and the difficulties created by the mixture of national and international impacts of transport activities.

The topic rightly has invited interdisciplinary treatment and a variety of approaches. An important subset of the approaches used involves economic analysis; in common with other economic literature these works tend to emphasize the role of markets in tracing the many effects, anticipated and otherwise, of policies. Economic approaches often consider pricing policies and attempt to evaluate their effectiveness in comparison with more traditional measures such as 'command and control' regulation and directed technological innovation. Another important subset of approaches involves simulation modelling, in which key relationships are represented mathematically so that their influence can be quantified and their inter-relationships discerned precisely. Such models are increasingly required by regulatory agencies as part of their policy development process, and in some cases (such as the US Clean Air Act) as part of the process of regulatory approval.

This book treats the intersection of these two subsets of approaches: simulation models with a strong economic content. This intersection defines a broad but powerful way to study environment and transport. Its breadth is illustrated by the wide range of policies, from carbon taxes to speed limits, treated here. Its power comes from putting insights about inter-related actions and the role of markets – the strong points of economic theory – into a form suitable for making quantitative predictions about the results of policies.

R. Roson and K.A. Small (eds.), Environment and Transport in Economic Modelling, 1–9.

In this introduction, we first describe in more detail the particular methods of analysis used in the book. We then discuss some broad themes that seem to emerge from the papers, and from other evidence, concerning the changes that can be expected from policies toward environment and transport. These observations suggest that environmental policies, even when highly successful in their objectives, have smaller effects on broad transport trends than usually believed; and that more significant changes in the transport sector are likely to result from addressing other objectives such as congestion and overall efficiency.

1.1. Contributions

All contributions included in this book illustrate real-world problems analysed using applied economic models. Most of them were developed from preliminary presentations at the international workshop, *Environment and Transport in Economic Modelling*, held in Venice in November 1995. The workshop was organised jointly by three organizations: Ca' Foscari University, the Italian National Research Council (PFT2 project) and the Fondazione ENI 'Enrico Mattei' in Milan. Purely theoretical models are not considered in this book, nor are contributions dealing solely with issues of measurement such as valuation of external costs; for these important topics a large and growing literature is available, addressing issues that are not specifically related to the transport sector.

Despite differences in perspective and methodology, the models used by our authors often share certain traits which make classification useful. Some authors, for example, use partial equilibrium models, focusing on just one part of the economy. This is especially appropriate when a fine level of detail is needed such as in the specification of alternative technologies, and when more complexity is not expected to add much to the qualitative picture of results. Other authors use general equilibrium models, which are needed to deal fully with systemic effects of transport and environmental policies. As demand for transportation services originates from all production and consumption activities, 'micro' measures affecting the transport sector may have important 'macro' consequences for the whole economic system.

Another useful distinction is between normative models which find optimal policies in the presence of given objectives, and positive models which assess the impact of specific policy actions. Normative models can most easily be found in research works following the public economics tradition, in which the concepts of externality and market failure have been elaborated. Traditional transportation models, on the other hand, are usually positive and lend themselves to 'what if' simulation exercises; they are better suited to specify transport system characteristics in detail. Sometimes the distinction between normative and positive models is blurred. For instance, when normative models are

Table 1.1. Key characteristics of the models

Authors	General or Partial Equilibr.?	Normative or Positive?	Distributional Implications? (Yes, No)	Urban or Other Scale?	Price or Non-price Instruments?	Country
DeBorger–Swysen	P	N	N	O	P	Belgium
Roson	P	N	Y	U	P	Italy
Peirson–Vickerman	P	N	N	U+O	P	UK
Harrison–Kristrom	G	P	Y	O	P	Sweden
Pireddu	G	P	Y	O	P	–
Harrington et al.	P	P	Y	O	P+N	USA
Bates	P	P	N	U	P	UK
Rietveld et al.	P	N	N	U+O	N	Netherlands

specified as programming problems, constraints can be added to represent a restricted policy menu. Also, using simulation models of a positive type, simulation exercises can be defined to determine which policy package is expected to produce the best results with respect to some criterion.

The interactions among policy objectives is also important. Factors like income distribution, political feasibility and international cooperation may be critical for the success of environmental policies. Models may take into account the existence of other policy objectives explicitly, for example by including additional variables in the optimisation program or by evaluating *ex post* the consequences of policy options for indicators of these other objectives.

Of course, the model structure depends critically on the geographical scale under consideration. This affects the type of environmental problems considered, because environmental spillovers may occur at local, regional, national, or even global scales. It is particularly useful to distinguish between models of a single urban area and models of larger regions because they usually address quite different issues. Traffic congestion and land use, for example, are typical urban problems, whereas economies of scale and market structure may become more important at a coarser geographical scale.

Policies in this field may be based on price instruments like taxes and subsidies, or on non-price instruments such as required maintenance programs, speed limits, and safety and other technical requirements. Although different instruments may have similar economic consequences, the political and technical feasibility of alternative instruments may differ greatly.

Models of all the different classes mentioned above can be found in the chapters of this book. Table 1.1 helps to pinpoint the key emphases of the models presented. Parameters have been estimated in all models, with the exception of Pireddu's, using a combination of calibration techniques (for structural parameters) and econometric techniques (mainly for demand/supply elasticities). In some cases, uncertainty about correct parameter values is accommodated through a sensitivity analysis.

Bruno De Borger and Didier Swysen introduce a model (TRENEN) designed to identify optimal fiscal policies in the transport sector. An unregulated market for transport services may fail in achieving allocative efficiency because several external costs (pollution, congestion, accidents) are not fully internalized by the agents. In this case, taxes and subsidies may be used to 'correct' market prices. In general, optimal tax levels are difficult to determine in this context, because marginal costs are not constant and several alternative transport technologies, with quite different impacts on the environment, are available. In the TRENEN model one or more representative consumers make a choice between alternative technologies on the basis of generalized transport costs, possibly including taxes or subsidies. In addition, indirect demand for transportation comes from the consumption of goods, whose production process requires inputs from the transport industry. The authors apply the model to Belgium.

The same model in a different version is used by *Roberto Roson* to estimate the optimal price structure for urban transportation services in the city of Bologna. The model does not predict higher consumption levels for public transport modes even though prices incorporate marginal social costs including environmental damage. The reason is that the current price structure is distorted by a variety of taxes and subsidies, and when they are eliminated the use of private automobiles goes up. Nevertheless, air emissions go down due to switches to cleaner cars. After briefly describing the institutional background of these existing market distortions, Roson considers what policy objectives could justify their existence. In particular, the potential role of distortionary taxation as an instrument of income redistribution is assessed. The model is then extended by including distributional objectives in the social welfare function, whose parameters are estimated from current policies.

John Peirson and Roger Vickerman consider optimal pricing of transport services from a different perspective. In addition to external costs, bringing about the introduction of Pigovian taxes, they consider the possible existence of large economies of scale and natural monopolies, supporting the introduction of subsidies. In addition, they distinguish between short run, in which the infrastructure is given, and long run, in which the infrastructure can be optimally adjusted. To this purpose, they use a multi-modal calibrated model which is applied to three cases: peak and off-peak traffic in London and inter-urban travel throughout the United Kingdom. As in the TRENEN model, they find that efficient pricing does not necessarily result in substantial shifts in demand to modes of transport with lower external costs. They also find, however, that results may be substantially different when large economies of scale are present and the infrastructure stock can be adjusted. Unfortunately, much uncertainty exists about the actual magnitude of economies of scale for the different transport modes. This uncertainty is accommodated in the paper through a sensitivity analysis.

Glenn Harrison and Bengt Kriström present an applied general equilibrium model of Sweden which is used to simulate the introduction of carbon, petrol and diesel taxes. The model identifies 87 industries, six types of labour and 30 classes of households, differentiated by family status and income. This high level of disaggregation makes it possible to trace in detail the distributional consequences of alternative fiscal policies. The general equilibrium structure of the model makes it also possible to compute the overall social costs of additional indirect taxation, and to make broad comparisons between those costs and the benefits of reduced pollution.

Giancarlo Pireddu uses a general equilibrium model which is more aggregated but has a richer structure for the treatment of externalities. Three production sectors are considered: a polluting sector, a polluted sector and a pollution abatement sector. The level of pollution abatement is endogenous, and the abatement technology has decreasing returns to scale. Environmental externalities affect the economic system by reducing consumers' utility and factor productivity, but also by inducing 'defensive' expenditure by firms and households. This expenditure includes all expenses incurred to avoid welfare or profit losses imposed by environmental externalities. Welfare and efficiency of alternative fiscal policies depend on how the tax revenue is redistributed and how the preexisting distortionary taxation is modified.

Winston Harrington, Virginia McConnell and Anna Alberini provide an interesting example of policy analysis in which price instruments are combined with regulations in the control of vehicle emissions. These authors consider the case of 'inspection and maintenance' programmes which complement emission standards for new vehicles in the USA. Two policy options are considered: a 'command and control' regimen, in which all cars not passing efficiency tests must be repaired, and a flexible maintenance regimen, in which car owners are free to choose whether and when to repair but a set of emission fees is introduced. Emission fees affect the incentives of car owners and indirectly influence the rate of repairing. A distinguishing feature of the model is the explicit treatment of uncertainty. Stochastic elements of the model include the initial emissions level, errors in emissions measurement, the effectiveness of vehicle repairs, and the ability of mechanics to predict repair effectiveness. Distributional implications are taken into account via the introduction of ceilings on total expenditure for repairs, repair subsidies and fee baselines.

John Bates illustrates the findings of a complex transport planning model for London known as APRIL. This model uses travel-time and price elasticities, derived from travel demand studies, to predict the responses of travellers to various scenarios characterized by travel times and prices for specific trips. It is primarily intended to assess the effects of a 'congestion charging' programme. The model output includes information on economic and environmental effects of various implementation schemes. This contribution illustrates how a 'traditional' network equilibrium model can be fruitfully extended to analyse environmental issues. The results suggest that raising the congestion charge beyond approximately US$6 for a typical one-way commute to Central London exhibits markedly diminishing returns in terms of congestion relief, but continuing benefits in terms of reduction in air emissions. The model is also used to examine alternative forms of geographical coverage of the congestion charging.

Piet Rietveld, Arjan van Binsbergen, Theo Schoemaker and Paul Peeters consider the possibility of optimal regulation of vehicle speeds for various types of road. Vehicle speeds are strongly correlated with various types of internal and external costs: pollution, accidents, energy consumption and others. Speed limits could be optimally set, with little cost, to balance social costs and benefits. The authors are aware that speed limits only indirectly influence actual speed, and that externalities also depend on the type of road. Using price elasticity parameters, they estimate how demand levels for the alternative road types change when speed limits are optimally set. The optimal speed limit is based on a trade-off between travel time (decreasing with speed) and total external and internal cost (increasing with speed). The case for the imposition of lower limits turns out to be strongest for roads with the highest speeds, namely highways and expressways.

1.2. Themes and policy context

It would be rash to claim that any overall conclusion is specifically supported by all the diverse models in this book. Nevertheless, a few themes come through often enough that they seem reasonable candidates for generalization, particularly because they are supported by other evidence. We identify four here.

First, environmental taxes do not cause large shifts in total travel by private vehicles. Several papers find that although pricing measures aimed at environmental externalities are quite successful in improving the environmental indicator targeted, their effectiveness does not depend mainly on reducing automobile travel. Rather, the main mechanisms are increased adoption of cleaner technology such as cars with catalytic converters or with better fuel efficiency. In fact, when environmental taxes are combined with other measures to improve efficiency, the effects of the latter are dominant and may increase motor vehicle use. Thus we have the apparent paradox in Roson's paper of efficient taxation resulting in more vehicles but less pollution.

The underlying reasons for this are explained by Small (1991) and Small and Gómez-Ibáñez (1998b). Basically the measurable externality costs associated with automobile travel are simply not large enough to make much of a change in the total cost of driving. Consider, for example, the air pollution cost estimates listed in Peirson and Vickerman's article in this volume, which are toward the high end of the range of estimates in the literature. The figure given for peak driving in London is only 15% of the estimated efficient total price for such driving; for inter-urban travel it is just 5%. The external costs of noise, accidents, and global warming are smaller still by these estimates. Small and Kazimi (1995) estimate the air pollution costs for the average car in the Los Angeles region in 1992 to be about US$0.02/km, again relatively minor in comparison to other costs of driving. They also find that accounting for a shadow value of carbon dioxide based on a stringent reduction path – even one that is far more stringent than any likely to be adopted through international negotiations on global warming – would add only another $0.02/km. Therefore, it is not plausible that charging for such external costs would cause many people to drastically curtail their use of automobiles.

Why, one might ask, are people so insistent on driving even in the face of high costs? The answers go beyond such simplistic notions as Californians' supposed love affair with the automobile or oil companies' conspiracies to choke off public transportation. Rather, the answers are found in the rising importance of time and flexibility in highly productive and affluent societies. Flexible transportation is increasingly an integral component of modern economic development with its emphasis on information, adaptability and competitive response to rapid change. Peoples' transportation needs are not routine but rather varied and unpredictable: precisely the kinds of trips catered to by the automotive transportation system. This is especially true in urban areas, which have always thrived on the innovative edge gained through mutual interaction.

The effect of environmental taxes on specific uses of technology, however, is another story. The papers by Roson and by De Borger and Swysen both illustrate how pollution charges can increase the adoption of catalytic converters. Furthermore, the work by Harrington, McConnell and Alberini shows that well-designed charges induce significant improvements in maintenance of the pollution control equipment, resulting in better emissions characteristics from whatever vehicle fleet is on the road. The evidence is that environmental costs, while small in comparison to total costs of driving, are large enough to have a considerable impact on choices of technology or of maintenance strategy. Thus environmental taxes achieve their objectives by leading people and businesses to seek targeted behavioural or technological changes that reduce adverse environmental impacts, rather than by satisfying well-intentioned but naive notions about the role of private vehicles in our societies.

A second theme is that charging the social cost of congestion does, in contrast, create considerable shifts in motor vehicle use. These shifts are not uniform or widespread but rather specific in location and time. Peirson and Vickerman's estimates for London show the external costs of congestion to be nearly four times as large as those of air pollution during the peak period. Bates' paper shows explicitly how practical policies aimed at internalizing these costs result in large reductions in traffic flows to those areas targeted – central or inner London depending on the policy scenario. It appears that changes in mode and time of day are the most important components of these shifts. Similar evidence of the sizeable effects of congestion charges has been compiled by Harvey (1994) and Small and Gómez-Ibáñez (1998a).

Third, environmental and congestion policies do not coincide in their objectives; they are mutually reinforcing but only weakly so. It is true that policies aimed at environmental problems and those aimed at curtailing congestion both cause some decline in use of private motor vehicles, but the places and times are very different. Congestion pricing reduces vehicle use at peak times and in congested locations, but may increase vehicle use at other times and places. This is illustrated by De Borger and Swysen's finding that optimal pricing would cut peak car traffic in Belgium by 11%, but increase off-peak traffic by 7%. Pollution charges will tend to reduce automobile use across the board, but by too small an amount to significantly affect congestion.

Aside from the modest reduction in overall traffic that may result from congestion pricing, what about improved vehicle emissions due to smoother traffic flow? Even this factor is in some doubt. Guensler and Sperling (1994) show that emissions are highly complex functions of average vehicle speed and it is quite possible that some will decline and others increase as a result of congestion relief.

It remains true that those seeking environmental improvements and those seeking improved performance of the urban highway system during peak hours may find themselves natural political allies. The Bay Area Economic Forum, an organization of business and public officials in the San Francisco region, put both policies on the local agenda as a two-pronged attacked on transportation problems (Bay Area Economic Forum, 1990). This effort was strongly supported by some but not all environmental organizations, illustrating the divergent objectives of constituencies concerned with the current weaknesses of transport policies.

A final theme is this: the most successful policies aimed at environmental and transport objectives are those that operate on the key behavioural margins for those objectives. Policies aimed at curtailing urban air pollution need to target the activities to which air emissions are especially sensitive. For global warming caused by carbon dioxide, this implies a focus on overall fuel efficiency. For other emissions, it implies a focus on actual on-the-road emissions of both new and used vehicles, the importance of which is emphasized by Harrington et al. in this volume and also by Glazer et al. (1995). New-vehicle emissions control has been highly successful in curtailing pollution from automobiles, and could bring further improvements with stronger policies toward heavy vehicles. There is, however, increasing evidence that the focus on new vehicles is reaching diminishing returns in cost-effectiveness. Air emissions now seem to be dominated by a small percentage of vehicles whose pollution control equipment is no longer effective due to age, improper maintenance, or deliberate tampering. Thus policies aimed at such vehicles may offer the best hope for further improvements.

Similarly, policies aimed at congestion need to target the specific times and locations that contribute to congestion. Harvey (1994) illustrates through simulation models how such policies differ from those designed simply to curtail automobile travel in general. Similarly the analysis of real-world road pricing schemes by Small and Gómez-Ibáñez (1998a) shows that their effectiveness depends strongly on being able to track marginal externality cost at least to a rough degree. The original toll rings surrounding Singapore and several cities in Norway, for example, cannot be persuasively shown to offer net social benefits, although each succeeded in its specific goals – reducing peak congestion in central Singapore and raising money for road improvements in Norway. In the same way, Bates shows that social benefits from hypothetical congestion pricing in London would be considerably higher for policies that distinguish among traffic flows by direction and location than for a simpler single-cordon peak fee.

These four themes suggest that dreams of rolling back the phenomenal worldwide growth of automobile use are unlikely to be realized. Yet specific improvements in the performance of the highway-based transport system are possible, both with respect to its main function of providing fast transportation and with respect to its environmental consequences. The policies discussed in this book offer options for making these improvements happen.

References

Bay Area Economic Forum, 1990, Market-based solutions to the transportation crisis, San Francisco: Bay Area Economic Forum.

Glazer, A., D.B. Klein and C. Lave, 1995, Clean on paper, dirty on the road: troubles with California's smog check, Journal of Transport Economics and Policy, 29, 85–92.

Guensler, R. and D. Sperling, 1994, Congestion pricing and motor vehicle emissions: an initial review, in National Research Council, Committee for Study on Urban Transportation Congestion Pricing, Curbing gridlock: peak-period fees to relieve traffic congestion, vol. 2: Commissioned papers, Transportation Research Board Special Report 242, Washington, DC: National Academy Press, 356–379.

Harvey, G.W., 1994, Transportation pricing and travel behavior, in National Research Council, Committee for Study on Urban Transportation Congestion Pricing, Curbing gridlock: peak-period fees to relieve traffic congestion, vol. 2: Commissioned papers, Transportation Research Board Special Report 242, Washington, DC: National Academy Press, 89–114.

Small, Kenneth A., 1991, Transportation and the environment, in R. Thord, ed., The future of transportation and communication. Borlånge: Swedish National Road Administration, 217–230.

Small, K.A. and J.A. Gómez-Ibáñez, 1998a, Road pricing for congestion management: the transition from theory to policy, in K.J. Button and E.T. Verhoef, eds., Road pricing, traffic congestion and the environment: issues of efficiency and social feasibility, Cheltenham, UK: Edward Elgar (forthcoming).

Small, K.A. and J.A. Gómez-Ibáñez, 1998b, Urban transportation, in Handbook of applied urban economics, Amsterdam: North-Holland (forthcoming).

Small, K.A. and C. Kazimi, 1995, On the costs of air pollution from motor vehicles, Journal of Transport Economics and Policy, 29: 7–32.

CHAPTER 2

Optimal Pricing and Regulation of Transport Externalities: A Welfare Comparison of Some Policy Alternatives[1]

Bruno De Borger and Didier Swysen

2.1. Introduction

This paper presents a first effort to the study of optimal pricing policies and technology choices in inter-regional transport in Belgium on the basis of a disaggregate simulation model. The literature on optimal taxation in the presence of externalities (e.g., Sandmo, 1975; Wijkander, 1985; Oum and Thretheway, 1988; Bovenberg and van der Ploeg, 1994), adapted for the specific case of congestion-type externalities (Mayeres and Proost, 1996; De Borger et al., 1996), provides the theoretical basis for the simulation analysis. The specific treatment of congestion is necessary because, unlike many other external effects, congestion directly affects consumer demand for passenger transport and producer demand for freight transport. This results in complex feedback effects that have to be taken into account.

Unfortunately, although the theoretical models referred to above have substantially increased our understanding of the optimal pricing problem in the transport sector, their practical and policy implications remain somewhat limited. First, these models are obviously not designed to capture fully the heterogeneity of transport demand: various modes have to be considered for both passengers and freight, the congestion contributions widely differ according to the period of the day, gasoline and diesel do not generate the same external pollution effects, etc. Second, congestion-type externalities imply that a full welfare optimum may not be attainable by the usual price and tax instruments (such as fuel taxes), since these do not allow spatial or temporal discrimination. Additional instruments, such as a toll system, may be required. Moreover, pricing policies are not the only instrument available to the authorities. Regulatory policies with respect to technologies can be implemented

R. Roson and K.A. Small (eds.), Environment and Transport in Economic Modelling, 10–38.

(e.g. norms with respect to catalytic converters), infrastructure policies can be developed, etc.

To analyse optimal pricing policies that take account of these complications, a simulation model that allows incorporation of the relevant heterogeneity of transport services may be useful. In this paper, we describe the development of such a model and present some preliminary empirical results. The model looks for optimal prices (or taxes) and supply characteristics of the different transport services. A sufficient degree of heterogeneity of services is allowed by using nested utility and cost functions. The model is a standard welfare optimization problem subject to relevant constraints on the policy instruments: it incorporates passenger and commodity transport, it takes account of all major external costs of the various transport modes, it captures the budgetary implications of government policies, and, finally, it allows general equilibrium effects of transport prices on other goods in the economy.

The model is very detailed in terms of transport services and externalities taken into account. To accomplish this some strong assumptions were made concerning other characteristics of the model. For example, location is assumed to be exogenously given, the road network is aggregated in one link, and the model is static in the sense that no explicit time dimension is included.

2.2. The theoretical model

In this section we present the theoretical structure of the simulation model used to study optimal transport pricing and regulation. The model is in the tradition of previous optimal taxation models with externalities, but its specific characteristics are tailored towards the transport industry. A brief overview of the structure of the model is as follows. There are two production sectors in the economy: a private sector, producing an aggregate consumption good, and a transport sector. The transport industry provides both final goods to consumers (passenger transport) and intermediate goods to private producers (freight transport). The latter are used as inputs in the production process of final goods. The household sector is modelled by assuming a representative household; in other words, although this would be desirable from a policy viewpoint, the current version of the model ignores distributional considerations. The government is assumed to be interested in maximizing welfare, using public transport prices and taxes to be applied to private transport services as instruments. Moreover, it can impose norms on technologies (for example with respect to catalytic converters, airbags, etc.) to improve welfare. The objective function takes account of budgetary implications of tax and pricing policies, and it captures the impact of all important transport externalities.

Transport services produce two types of externalities. The first consists of external effects that not only directly affect consumer welfare but also have an impact on demand behaviour. As previously suggested, congestion is the most obvious example. High traffic levels cause travel speed to decrease, and this directly affects the demands for the various transport modes. The second type of externality captures effects which certainly affect consumer utility, but probably do not influence demand behaviour. Air pollution provides a good example. For most people, air pollution affects utility, but it is not an important determinant of their travel demand.

2.2.1. *The behaviour of households*

A representative household maximizes utility subject to a budget restriction. The demand for transport service i (expressed in passenger-km) is denoted by X_i^p $(i = 1, \ldots, I)$. Other goods are aggregated in a composite commodity X. It is assumed that consumers take the prices of passenger transport q_p^i, the price of 'other goods', q, and congestion C as exogenously given. Consumer income is denoted by R. Solving the corresponding utility maximization problem yields the demand functions for passenger-km with the various modes, the demand for other goods, and the indirect utility function V.

$$X_i^p = X_i^p(q_1^p, \ldots, q_I^p, q, C, R) \quad \forall\, i$$

$$X = X(q_1^p, \ldots, q_I^p, q, C, R)$$

$$V = V(q_1^p, \ldots, q_I^p, q, C, R)$$

2.2.2. *The private sector*

The private sector consists of a large number of competitive firms. Without loss of generality, these firms are aggregated. This aggregate private sector produces the final goods X according to a constant returns to scale technology using, among others, various freight transportation services as inputs. The number of ton-km with transport service (e.g., mode) j is denoted X_j^f $(j = 1, \ldots, J)$. Other inputs are aggregated in a composite input denoted X^o. It is assumed that congestion imposes a negative production externality, and that, for each level of X demanded by consumers, the firms minimize total cost. This implies the system of input demand functions

$$X_j^f = X_j^f(q_1^f, \ldots, q_J^f, q^o, C, X) \quad \forall\, j$$

$$X^o = X^o(q_1^f, \ldots, q_J^f, q^o, C, X)$$

where q_j^f and q^o are the prices for freight transport and other inputs, respectively. Constant returns to scale in production combined with marginal cost pricing implies that output is demand-determined,[2] and that the equilibrium price q of the final good can be written as

$$q = q(q_1^f, \ldots, q_J^f, q^o, C)$$

2.2.3. *The transport sector*

The transport sector produces passenger- and ton-km using a large variety of transport alternatives. Moreover, on the supply side a number of different technologies are available. The number of passenger-km (using alternative i) supplied with technology k *is denoted by* $Z^p_{i,k}$ $(k = 1, \ldots, K)$. An example may be instructive. For instance, let X^p_1 be the demand for passenger-km with alternative 1, say large cars with heavy fuel consumption. Then $Z^p_{1,1}$ could be the supply of passenger-km produced with cars of this type that are not equipped with a catalytic converter, and $Z^p_{1,2}$ could denote the corresponding number of passenger-km of fuel-demanding cars that are equipped with a catalytic converter. The private cost of providing one pass-km for mode i with technology k is denoted $c^p_{i,k}$ and is assumed to be constant. The externality cost (other than external costs associated with congestion) of one pass-km with mode i and technology k is denoted $e^p_{i,k}$ and also assumed to be constant.[3] Similarly, let X^f_j denote the demand for ton-km with freight transport alternative j. Then $Z^f_{j,r}$, $c^f_{j,r}$ and $e^f_{ej,r}$ are the supply of ton-km with freight transport alternative j using technology r, and the corresponding unit (private) cost and external cost, respectively. Obviously we have

$$X^p_i = \sum_{k=1}^{K} Z^p_{i,k} \quad \forall i$$

$$X^f_j = \sum_{r=1}^{R} Z^f_{j,r} \quad \forall j$$

Note that the choice of emission technology will also be determined in an optimal way by the authorities. Consider, for example, emission technologies for cars. Given the public good nature of cleaner technologies this would result in underprovision of emission reducing technologies. We therefore opted for an alternative approach: emission technology to be provided by suppliers will result from the overall optimization and will be such as to minimize social costs.

2.2.4. *The nature of congestion*

At its most general level, congestion is specified as a function of the use of all freight and passenger transport alternatives[4]

$$C = g(X^p_1, \ldots, X^p_I, X^f_1, \ldots, X^f_J)$$

As previously suggested, congestion differs from other external costs in that it explicitly enters all consumer demand functions as well as the production function in the private sector. This implies that changing transport prices generate complex reactions in congestion and demand. Consider for instance the effect of an increase in the price of the ith passenger transport mode, q^p_i.

Using the respective specifications of the demand functions and the definition of congestion, it is straightforward to show that the ultimate impact on congestion is given by

$$\frac{dC}{dq_i^p} = \frac{\sum_{n=1}^{I} \frac{\delta g}{\delta X_n^p} \frac{\delta X_n^p}{\delta q_i^p} + \sum_{m=1}^{J} \frac{\delta g}{\delta X_m^f} \frac{\delta X_m^f}{\delta X} \frac{\delta X}{\delta q_i^p}}{1 - \eta}$$

where

$$\eta = \sum_{n=1}^{I} \frac{\delta g}{\delta X_n^p} \left(\frac{\delta X_n^p}{\delta q} \frac{\delta q}{\delta C} + \frac{\delta X_n^p}{\delta C} \right) + \sum_{m=1}^{J} \frac{\delta g}{\delta X_m^f} \left(\frac{\delta X_m^f}{\delta C} + \frac{\delta X_m^f}{\delta X} \left(\frac{\delta X}{\delta q} \frac{\delta q}{\delta C} + \frac{\delta X}{\delta C} \right) \right)$$

The numerator of the above expression measures the direct impact of the price increase on the level of congestion, both via passenger demand and freight demand. Changes in freight traffic demand are indirectly induced by changes in private good demand that lead to adjustments in production levels and, therefore, in the demand for inputs. The denominator of the above expression corrects the direct effect for the feedback effect. The change in the level of congestion C itself affects the demand for a variety of transport services. The feedback effect is typically negative so that the denominator is greater than one: the overall impact of a price change on congestion is smaller than the direct effect.

Similarly, the total impact on congestion of a change in the price of freight transport mode j can be written as

$$\frac{dC}{dq_j^f} = \frac{\sum_{n=1}^{I} \frac{\delta g}{\delta X_n^p} \frac{\delta X_n^p}{\delta q} \frac{\delta q}{\delta q_j^f} + \sum_{m=1}^{J} \left(\frac{\delta g}{\delta X_m^f} \left(\frac{\delta X_m^f}{\delta q_j^f} + \frac{\delta X_m^f}{\delta X} \frac{\delta X}{\delta q} \frac{\delta q}{\delta q_j^f} \right) \right)}{1 - \eta}$$

Not surprisingly, freight rates affect congestion through two distinct channels: directly through the demand for freight transport, and indirectly through their impact on private good prices.

2.2.5. *The planning problem*

The planner is assumed to choose all prices and the technologies to be implemented so as to maximize the following problem

$$\max_{q_i^p, q_j^f, Z^p_{i,k}, Z^f_{j,k}} W = \frac{1}{\mu_0} V(q_1^p, \ldots, q_I^p, q(q_1^f, \ldots, q_J^f, C), C, R)$$

$$+ (1 = \lambda)\left(\sum_{i=1}^{I} \sum_{k=1}^{K} (q_i^p - c^p_{i,k}) Z^p_{i,k} + \sum_{j=1}^{J} \sum_{r=1}^{R} (q_j^f - c^f_{j,r}) Z^f_{j,r} \right)$$

$$- \left(\sum_{i=1}^{I} \sum_{k=1}^{K} e^p_{i,k} Z^p_{i,k} + \sum_{j=1}^{J} \sum_{r=1}^{R} e^f_{j,r} Z^f_{j,r} \right)$$

s.t.

$$\sum_{k=1}^{K} Z^p_{i,k} = X_i^p(q_1^p, \ldots, q_I^p, q(q_1^f, \ldots, q_J^f, C), C, R) \quad \forall i$$

$$\sum_{r=1}^{R} Z^f_{j,r} = X_j^f(q_1^f, \ldots, q_J^f, C, X(q_1^p, \ldots, q_I^p, q(q_1^f, \ldots, q_J^f, C), C, R)) \ \forall j$$

The objective function consists of three terms. The first term measures the representative consumer's indirect utility, normalized by the marginal utility of income in a reference situation so as to reflect consumer welfare in terms of real income. The second term measures tax revenue weighted by one plus the shadow cost of public funds.[5] The final term gives the monetary value of all external effects (e.g., environmental damage) other than congestion. Of course, the welfare effects of congestion are directly captured in the consumer's indirect utility function. Finally, the constraints indicate that for each transport alternative the sum of services produced by all available technologies should equal the corresponding demand for this particular service.

2.2.6. *Optimal taxation rules*

To focus on the design of optimal taxes, suppose initially that only one technology is available for each transport service. In that case, we can rewrite the objective function as:

$$\max_{\substack{q_i^p, q_j^f \\ \forall i,j,k}} = \frac{1}{\mu_0} V(q_1^p, \ldots, q_I^p, q(q_1^f, \ldots, q_J^f, C), C, R)$$

$$+ (1 + \lambda)\left(\sum_{i=1}^{I} (q_i^p - c_i^p) X_i^p + \sum_{j=1}^{J} (q_j^f - c_j^f) X_j^f \right)$$

$$- \left(\sum_{i=1}^{I} e^p_{i,k} X_i^p + \sum_{j=1}^{J} e_j^f X_j^f \right)$$

The first-order conditions with respect to the prices of passenger mode i (q_i^p) and with respect to freight transport mode j (q_j^f) can be rearranged so as to yield

$$\sum_{n=1}^{I}\left(\frac{q_n^p - c_{n,k}^p - \dfrac{e_{n,k}^p + \left(\dfrac{\text{MSDC}}{1-\eta}\right)\dfrac{\delta C}{\delta X_n^p}}{1+\lambda}}{q_n^p}\right)\varepsilon_{X_n^p, q_i^p}\frac{X_n^p q_n^p}{X_i^p q_i^p}$$

$$+\sum_{m=1}^{J}\left(\frac{q_m^f - c_{m,r}^f - \dfrac{e_{m,r}^f + \left(\dfrac{\text{MSDC}}{1-\eta}\right)\dfrac{\delta C}{\delta X_m^f}}{1+\lambda}}{q_m^f}\right)\varepsilon_{X_m^f X}\varepsilon_{X, q_i^p}\frac{X_m^f}{X_i^p}\frac{q_m^f}{q_i^p}$$

$$=\frac{\dfrac{\mu}{\mu_0}-(1+\lambda)}{1+\lambda}\qquad \forall i$$

$$\sum_{n=1}^{I}\left(\frac{q_n^p - c_n^p - \dfrac{e_n^p + \left(\dfrac{\text{MSDC}}{1-\eta}\right)\dfrac{\delta C}{\delta X_n^p}}{1+\lambda}}{q_n^p}\right)\varepsilon_{X_n^p, q}\varepsilon_{q, q_j^f}\frac{X_n^p q_n^p}{X_j^f q_j^f}$$

$$+\sum_{m=1}^{J}\left(\frac{q_m^f - c_m^f - \dfrac{e_m^f + \left(\dfrac{\text{MSDC}}{1-\eta}\right)\dfrac{\delta C}{\delta X_m^f}}{1+\lambda}}{q_m^f}\right)\left(\varepsilon_{X_m^f, q_j^f} + \varepsilon_{X_m^f, X}\varepsilon_{X, q}\varepsilon_{q, q_j^f}\right)\frac{X_m^f q_m^f}{X_j^f q_j^f}$$

$$=\frac{\dfrac{\mu}{\mu_0}-(1+\lambda)}{1+\lambda}\qquad \forall j$$

In these expressions μ is the marginal utility of income, and $\varepsilon_{z,t}$ denotes the elasticity of z with respect to t. Optimal pricing rules depend on own and cross-price elasticities of demand for passenger and freight transport and for the private good, on the output elasticity of freight transport demand, and on the elasticity of private goods prices with respect to freight rates. Finally, MSDC is the marginal social damage of congestion; it corresponds to the full external cost of an increase in congestion and is given by

$$\begin{aligned}
\text{MSDC} = &-\left(\frac{1}{\mu_0}\frac{\delta V}{\delta C}\right) - \left(\frac{1}{\mu_0}\frac{\delta V}{\delta q}\frac{\delta q}{\delta C}\right) \\
&-(1+\lambda)\left(\sum_{i=1}^{I}(q_i^p - c_i^p)\left(\frac{\delta X_i^p}{\delta q}\frac{\delta q}{\delta C} + \frac{\delta X_i^p}{\delta C}\right)\right. \\
&\left.+\sum_{j=1}^{J}(q_j^f - c_j^f)\left(\frac{\delta X_j^f}{\delta C} + \frac{\delta X_j^f}{\delta X}\left(\frac{\delta X}{\delta q}\frac{\delta q}{\delta C} + \frac{\delta X}{\delta C}\right)\right)\right) \\
&+\left(\sum_{i=1}^{I} e_i^p\left(\frac{\delta X_i^p}{\delta q}\frac{\delta q}{\delta C} + \frac{\delta X_i^p}{\delta C}\right)\right. \\
&\left.+\sum_{j=1}^{J} e_j^f\left(\frac{\delta X_j^f}{\delta C} + \frac{\delta X_j^f}{\delta X}\left(\frac{\delta X}{\delta q}\frac{\delta q}{\delta C} + \frac{\delta X}{\delta C}\right)\right)\right)
\end{aligned}$$

It clearly consists of four identifiable terms. The first one is the direct effect of an increase in congestion on the consumers' utility. Note that this effect is expressed in monetary units by dividing by the marginal utility of income. The second term represents the welfare cost of an increase in the price of final goods, induced by a change in congestion level. The third term measures the impact of congestion on the government budget, multiplied by one plus the shadow cost of public funds. The fourth term stands for the changes in externalities other than congestion due to increases in congestion.

Not surprisingly, the optimality rules are quite complex and provide little scope for direct interpretation. However, it is easy to show that under suitable restrictions they reduce to simple rules well-known from the literature. Suppose, for example, that the approximation of the marginal utility of income is perfect (i.e., $\mu = \mu_0$), and let all cross-price elasticities of demand for both passenger and freight transport be zero.[6] We find under those assumptions

$$\left(\frac{q_i^p - c_i^p - \dfrac{e_i^p + \left(\dfrac{\text{MSDC}}{1-\eta}\right)\dfrac{\delta C}{\delta X_i^p}}{1+\lambda}}{q_i^p}\right) = -\frac{\lambda}{(1+\lambda)\varepsilon_{X_i^p, q_i^q}} \quad \forall i$$

$$\left(\frac{q_j^f - c_j^f - \dfrac{e_j^f + \left(\dfrac{\text{MSDC}}{1-\eta}\right)\dfrac{\delta C}{\delta X_j^f}}{1+\lambda}}{q_j^f}\right) = -\frac{\lambda}{(1+\lambda)\varepsilon_{X_j^f, q_j^f}} \quad \forall j$$

If perfect tax instruments were available, i.e., $\lambda = 0$, these expressions simply boil down to marginal social cost pricing. If no lump-sum taxes can be used

($\lambda > 0$) then the 'mark-up' of price over marginal external cost varies inversely with the demand elasticity. However, the mark-up is not over social marginal cost nor over private marginal cost. It is a mark-up over private marginal cost plus a fraction of marginal externality cost. This result is well-known (e.g., Sandmo, 1975; Bovenberg and van der Ploeg, 1994; Oum and Tretheway, 1988).

2.2.7. *Optimal choice of technologies*

We now briefly consider the more general problem given before, in which not only taxes but also the choice of technologies can be used as policy instruments. It is easy to show that optimal taxation rules are identical to those presented in the previous subsection. With respect to the optimal choice of technology, note that, if the private and external costs were non-constant and lump-sum taxes were available ($\lambda = 0$), the first-order conditions with respect to $Z^p_{i,k}$ and $Z^f_{j,r}$ would require that the various technologies be implemented up to the level where all their respective relevant marginal social costs were equalized (see De Borger and Swysen, 1996 for details). When $\lambda > 0$, we would also have to consider the welfare effects of the tax revenues that are foregone by having a higher resource cost.

However, if one assumes that the marginal private and external cost of providing one passenger-km or ton-km with mode i and technology k is constant, then it is clear that we obtain a corner solution. Optimally only one technology will be supplied for passenger transport and one for freight transport, viz. those technologies that produce the optimal traffic volumes at lowest social costs will be introduced. This of course implies that a technology with a higher resource cost can be implemented if it has comparatively lower external costs.

2.3. The simulation model

In this section we describe the construction of the simulation model used to determine optimal pricing policies for a large number of transport alternatives. However, before turning to the formulation of the model it is instructive to briefly review its most important limitations. First, it is a static model in the sense that the localization of households and firms is assumed to be exogenously given. Second, the model provides a representation of an equilibrium situation with a fully adapted stock of transport means. This implies that automobile ownership is not explicitly treated as an independent variable. Although car ownership is endogenous, we use a reduced-form model of mode choice, applicable to a time frame long enough for car ownership to adjust to changes in other variables. After a policy change, the model calculates the new equilibrium outcome, but dynamic adjustments are not explicitly described. The results of the simulations should be interpreted in a medium-term perspective, i.e. for a time horizon that is long enough to have fully adapted stock of transport

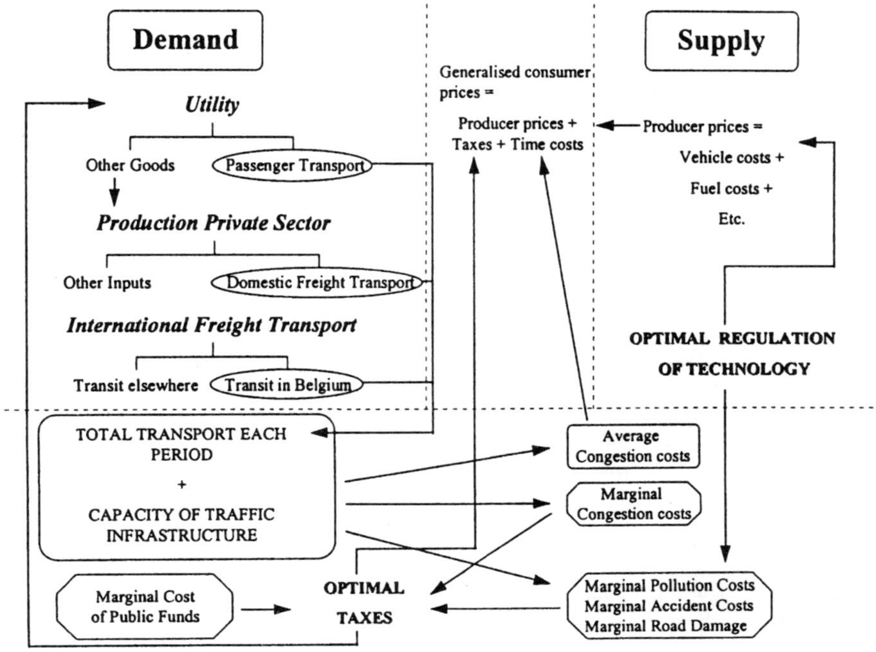

Figure 2.1. Stucture of the inter-regional model.

means, but not long enough to involve locational changes. Third, the model is not spatially disaggregated. In other words, transport is represented by one link per mode and there is no possibility of changing route.[7] Fourth, although this is desirable from a policy viewpoint, the version of the simulation model used in this paper does not yet capture neither the international dimensions of transport policies (e.g., the potential of tax exporting behaviour) nor their distributional implications.

The simulation model is sufficiently detailed so as to distinguish between the peak and off-peak periods of the day, it includes all relevant modes (for passengers: the private car, bus, and rail; for freight: truck, rail, inland waterways), it takes account of different types of cars as well as various fuel types (petrol, diesel). Moreover, the model captures most relevant externalities associated with transport services, including congestion, road surface depreciation, accident risks (i.e., safety), and various emissions. The latter include the contribution of transport to the greenhouse effect (CO_2-equivalent), to the ozone problem (VOC, NO_x), its contribution to acid rain (SO_2, NO_x), and to local air quality (CO, particulates).

The overall model structure is represented in Figure 2.1. (see also De Borger et al., 1995 for more details). We now briefly turn to the specification of the demand and supply sides of the transport market.

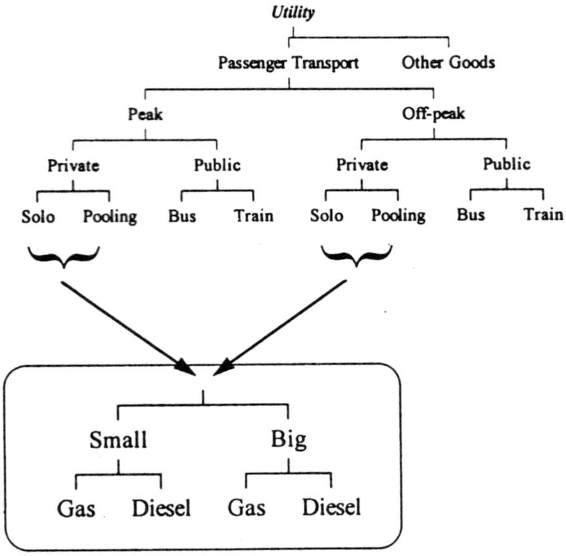

Figure 2.2. Multi-level decision structure for passenger transport.

2.3.1. *Structure of the demand side of the transport market*

Demand for passenger transport: consumers' behaviour

To structure transport demand decisions with a large number of alternatives we used nested CES-utility functions to represent consumer preferences (see Keller, 1976). The homothetic nature of the CES implies that, at each level, quantity and price indices can be constructed as functions of lower level quantities and prices, respectively. Moreover, each quantity index has a sub-utility interpretation. The CES-approach assumes that at each level subutility is separable in the different goods. Calibration of this function is based on estimates of substitution elasticities at the different levels combined with observations on prices and quantities at the lowest level of the tree structure.

The nested structure used for the simulation exercises is represented on Figure 2.2. At the highest level total utility depends on two aggregate goods, viz. transport and other goods. At the second level, the transport subutility component contains transport demands in two periods of the day (peak and off-peak) as arguments. At the third level, peak transport demand includes 'private' and 'public' peak demand. At the fourth level, public transport can be either bus or train. Furthermore, private transport (i.e car) consists of 'car pooling' and 'driving solo'. Car pooling is considered as a particular mode in order to allow different prices (per car-km) according to the car's occupancy rate. Furthermore, two car sizes are being considered, large and small. Finally, there are two possible fuels, petrol and diesel.

At each level aggregate price and quantity indices can be constructed. The aggregate quantity index of service i belonging to a given level n is given by

$$X_{n,i} = \left(\sum_{j \in i} \alpha_{i-1,j}^{\rho_{n,i}} \cdot X_{n-1,j}^{\rho_{n,i}} \right)^{1/\rho_{n,i}}$$

$$\text{where} \quad \rho_{n,i} = \frac{\sigma_{n,i} - 1}{\sigma_{n,i}}$$

The sigma refer to elasticities of substitution, and the notation $j \in i$ means that all disaggregates j are considered that branch off from alternative i at level $n-1$. The corresponding aggregate price index related to $X_{n,i}$ is given by

$$q_{n,i} = \left(\sum_{j \in i} \alpha_{n-1,j} \cdot q_{n-1,j}^{\tilde{\rho}_{n,i}} \right)^{1/\tilde{\rho}_{n,i}}$$

$$\text{where} \quad \tilde{\rho} = \frac{\tilde{\sigma}_{n,i} - 1}{\tilde{\sigma}_{n,i}}; \qquad \tilde{\sigma}_{n,i} = \sigma_{n,i}^{-1}$$

With the above structure, it is easy to show that the utility maximizing demand functions for commodity i can be written as

$$X_i = \frac{R}{q_{N,i}} \prod_{n=1}^{N} \alpha_{n-1,i} \left(\frac{q_{n,i}}{q_{n-1,i}} \right)^{\sigma_{n,i}}$$

where, as before, R is the consumer's income (see Keller, 1976). Note that the demand for an arbitrary good is a function of all prices via the aggregate price indices at the various levels.

In order to introduce congestion into the model, it will be assumed in the application that demands depend on generalized prices, defined as the sum of monetary and time expenditures. The demand for passenger-km with a given mode is assumed to depend on prices and speeds of all transport modes. Further details are given below.

The choice of the nested CES demand structure implies that the income elasticity equals 1; moreover, it imposes some restrictions on the substitution possibilities between goods in different branches of the tree. In particular, it implies that a price change in one branch will affect the demands for all goods in a given other branch in the same way. This implies, for example, that the elasticity of the demand for off-peak car transport and for off-peak public transport with respect to the price of peak bus transport are equal. Given these restrictions, the substitution elasticities were chosen such that the resulting price elasticities were close to those available in the literature (e.g. Oum et al., 1992; Goodwin, 1992). The price elasticities that were used are given in Table 2.1.[8]

Table 2.1. Passenger transport's price elasticities.

	Peak demand		Off-peak demand	
	Private	Public	Private	Public
Peak prices				
Private	−0.423	0.084	0.084	0.084
Public	0.003	−0.311	0.003	0.003
Off-peak prices				
Private	0.089	0.089	−0.512	0.089
Public	0.004	0.004	0.004	−0.369

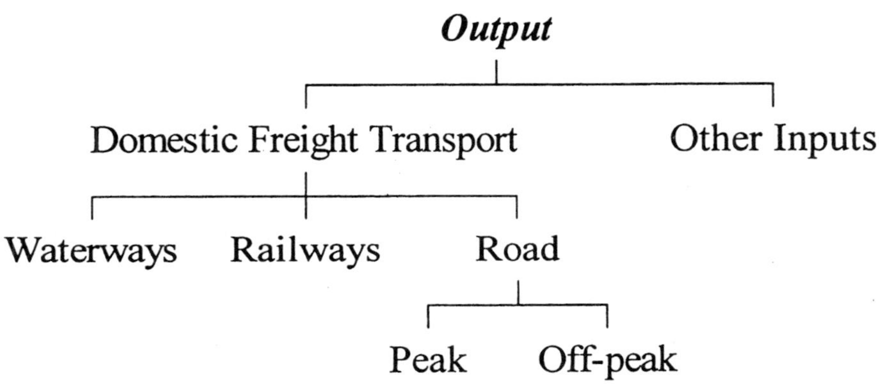

Figure 2.3. Multi-level decision structure for domestic freight transport.

Demand for freight transport

Freight transport demands are treated like derived demands for inputs by a private sector producing an aggregate private consumption good. This good enters the utility function of the representative consumer at the highest level. Consumer demand for this good generates production by the private sector, in which freight transport is one among several inputs. The demand for freight transportation is then assumed to be the result of cost-minimizing behaviour by producers, conditional on the output level of the final good to be produced.

Again, a tree structure is used to represent producers' decisions with a large number of transport alternatives. The tree considered in the simulation model is shown in Figure 2.3. It is developed in less detail than in the case of passenger transport, as the number of relevant alternatives is much smaller. For example, the peak versus off-peak decision is probably only relevant for road transportation and almost all road freight transport uses diesel. We therefore simply distinguish the three relevant modes (road, rail, inland waterways), and make a further distinction for road transport according to period of the day. No further refinements have been considered. Again, cost functions were assumed

Table 2.2. Freight transport's price elasticities.

	Demand road	Demand rail	Demand waterways
Price road	−0.561	0.017	0.017
Price rail	0.004	−1.456	0.004
Price waterways	0.002	0.002	−1.302

to be of the CES-type. In other words, for each level price and quantity indices are constructed along the same lines as for passenger transport. The price elasticities that were used are shown in Table 2.2. The substitution elasticities were choosen such that the resulting own-price elasticities for road and rail are close to those given by Oum et al. (1992). Note that here, as well generalized prices are being used (see below).

2.3.2. *Representation of supply*

The supply of the various transport modes is introduced in the model via cost functions for the different modes and alternatives. For freight transport and public passenger transport, resource costs include expenditures on labour, energy, materials, rolling stock etc. For private passenger transport (i.e. car) resource costs consist of depreciation expenditures, insurance, energy, parking costs, maintenance, etc.

As previously suggested, the choice of emission technology is also captured as part of the supply component. The emission technology to be provided by suppliers will result from the overall optimization and will be such as to minimize social costs. The decision variables here are the quantities supplied of vehicles equipped with a certain type of emission technology. Note, however, that in the simulation exercises reported below, choice between different technologies is only available for passenger transport by car.

2.3.3. *Externalities taken into account*

We considered the following external costs caused by transport: congestion, air pollution, accident risks, and road depreciation. First consider congestion. It was introduced in the model by specifying all demand functions for both passenger and freight transport in terms of generalized prices per kilometer, which include monetary expenditures as well as the monetary value of the time needed to travel 1 km. The monetary value of travel time/km is simply the value of time[9] per hour times the inverse of the speed in km/h. The generalized prices assume that speed of road traffic is endogenously determined according to a speed–flow relationship. This gives the speed of private car transport as a function of the number of passenger car units (PCU)[10]/h for each period considered. It was assumed that large and small cars have the same impact on congestion. Bus and truck speeds on the other hand were assumed to be a

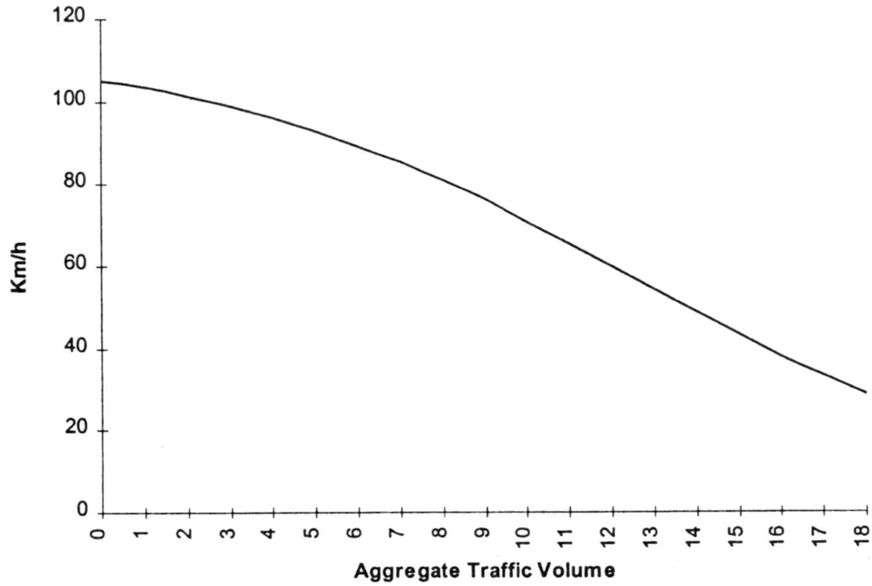

Figure 2.4. Speed–flow relationship.

constant fraction of cars' speed. Speeds of other modes of transport (rail, inland waterways) were assumed to be constant and independent from the overall level of traffic. The speed–flow relation was constructed by calibrating the parameters of the functional form suggested by Kirwan et all. (1995) on the basis of observable information on Belgian inter-regional speeds and flows. The use of aggregate data for a one-link network is reflected in the slope of the speed–flow relationship, which is graphically depicted in Figure 2.4.

All externalities other than congestion were assumed to have a constant cost/km. Air pollution is caused by vehicle emissions. We considered six pollutants: SO_2, NO_x, HC, CO, CO_2, and PM. Both the emission levels and the appropriate valuations were derived from available studies in the literature (refer to Mayeres, 1993 for more details on the valuation of CO_2, HC, SO_2 and NO_x, and to Small and Kazimi, 1995 for CO and PM). Accident costs were supposed constant per vehicle-km. They were derived using the methodology described in Mayeres et al. (1996); they take into account both material and physical damages. Finally, consistent with Newbery (1988), road depreciation was fully attributed to trucks.

2.4. Some simulation results

In this section, we illustrate the use of the model with Belgian data for 1991 on inter-regional transport flows and prices. The data concerning resource

Table 2.3. Reference situation 1991.

		Pass-km/ day (million)	Price (ECU/ pass-km)	Tax (ECU/ pass-km)	Marginal external cost	Speed (km/h)	Market share
Passenger transport							
Private transport							
Peak							
Large	Petrol	18.110	0.353	0.117	0.235		
	Diesel	11.572	0.282	0.082	0.231	55	40.9%
Small	Petrol	27.510	0.216	0.072	0.232		
	Diesel	16.867	0.180	0.054	0.227		
	Total	74.059					
Off-peak							
Large	Petrol	18.639	0.357	0.120	0.064		
	Diesel	11.910	0.285	0.084	0.060	95	42.1%
Small	Petrol	28.314	0.223	0.077	0.062		
	Diesel	17.360	0.186	0.058	0.057		
	Total	76.223					
Technology							
Large	Petrol			Standard			
	Diesel			Standard			
Small	Petrol			Standard			
	Diesel			Standard			
Public transport							
Peak							
Bus		5.462	0.030	−0.0117	0.0128	42.3	3.0%
Train		8.502	0.046	−0.0600	0.0003	70.0	4.7%
Total		13.964					
Off-peak							
Bus		6.503	0.030	0.0030	0.0055	73.1	3.6%
Train		10.123	0.046	−0.0084	0.0007	70.0	5.6%
Total		16.625					
Total pass-km		180.872					

Continued

costs, demands, prices, emissions, values of time, etc., were derived from a large number of sources. More details concerning the data are available in De Borger and Swysen (1996).

Solution of the model runs in two steps. First, using the available data we calibrate all remaining model parameters (utility function parameters, cost function parameters, ...).[11] The reference situation can then be interpreted as reflecting an initial market equilibrium consistent with observable information on prices and flows. The most relevant information with respect to the reference solution is presented in Table 2.3.[12] It suggests that all transport services are

Table 2.3. (*Continued.*)

Freight transport						
	Ton-km/ day (million)	Price (ECU/ ton-km)	Tax (ECU/ ton-km)	Marginal external cost	Speed (km/h)	Market share
Road						
Peak	14.026	0.058	0.0077	0.0403	35.7	67%
Off-peak	61.303	0.058	0.0077	0.0095	61.7	
Waterways	14.307	0.029	0.0035	0.0016	10.0	13%
Railways	22.337	0.038	0.0000	0.0012	55.0	20%
Total ton-km	111.973					

	Welfare components (in million ECU/day)
Utility	706.7488
Pollution passenger transport	1.56223
Pollution freight transport	0.24498
Accident costs	4.59839
Road damage	0.10598
Social welfare	700.2372

priced substantially below their corresponding marginal social costs in the peak period. In the off-peak, taxes and marginal external costs are quite similar, with the exception of large cars; the consumers actually pay more than the external costs. The reason is that in 1991 relatively little congestion existed on inter-regional traffic flows. Also note that, with the exception of bus transport in the off-peak period, public transport is heavily subsidized.

In a second step, once the model parameters have been determined we look for the values of the policy variables (prices, taxes, ...) that maximize the objective function. Importantly, note that we took the particular case of a zero shadow cost of public funds in the preliminary application reported here. This specification is equivalent to assuming lump-sum taxation. The objective function under these circumstances consists of just two components. First, it includes the indirect utility of the representative consumer, normalized by the marginal utility of income in the reference situation. Congestion is directly captured in utility via the generalized prices of transport services. It is assumed that tax revenues generated on the transport sector are lump-sum redistributed to the consumer. Tax income changes are thus explicitly taken care of in the welfare calculation via the representative consumer's indirect utility. A second component captures the external costs other than congestion.

To obtain some insight into the possible range of problems the model can handle we simulated several different applications of the model. The applications differ according to the pricing instruments one assumes the government has available to differentiate between prices of the various transport services,

Table 2.4. Optimal pricing.

		Pass-km/ day (million)	Price (ECU/ pass-km)	% change[a] (ECU/ pass-km)	Tax (ECU/ pass-km)	Marginal external cost	Speed (km/h)	Market share
Passenger transport								
Private transport								
Peak								
Large	Petrol	18.630	0.408	16%	0.172	0.172		
	Diesel	10.532	0.368	30%	0.168	0.168	63.7	38%
Small	Petrol	23.330	0.313	45%	0.169	0.169		
	Diesel	13.123	0.290	61%	0.165	0.164		
	Total	−11%						
Off-peak								
Large	Petrol	22.199	0.303	−15%	0.066	0.066		
	Diesel	12.756	0.262	−8%	0.061	0.061	94.2	47%
Small	Petrol	29.569	0.210	−6%	0.064	0.063		
	Diesel	16.768	0.187	1%	0.059	0.059		
	Total	7%						
Technology								
Large	Petrol				Standard			
	Diesel				Standard			
Small	Petrol				Standard			
	Diesel				Standard			
Public transport								
Peak								
Bus		6.412	0.051	71%	0.0094	0.0094	49	4%
Train		4.797	0.107	130%	0.0003	0.0003	70	3%
Total		−20%						
Off-peak								
Bus		6.680	0.032	9%	0.0056	0.0056	72.6	4%
Train		9.426	0.055	19%	0.0007	0.0007	70.0	5%
Total		−3%						
Total pass-km		−4%						

Continued

and according to whether or not one allows optimal technology choice. In a first exercise we studied optimal pricing under the assumption that the government has sufficient policy instruments (for example, vehicle taxes, fuel taxes, and tolls) available to allow optimal pricing of all transport alternatives considered.[13] However, we ignore in this first application the optimal choice of technologies. Second, we analysed a model that determines optimal prices when optimal taxes cannot be varied according to peak versus off-peak period because no tolls can be implemented. Again the optimal choice of technologies was ignored. Third, we illustrate the logic of the model with respect to the choice of optimal technologies using a very simple and specific example. Fourth,

Table 2.4. (*Continued.*)

Freight transport							
	Ton-km/ day (million)	Price (ECU/ ton-km)	% change[a] (ECU/ ton-km)	Tax (ECU/ ton-km)	Marginal external cost	Speed (km/h)	Market share
Road							
Peak	12.258	0.080	37%	0.0292	0.0292	41	65%
Off-peak	60.464	0.061	4%	0.0098	0.0098	61	
Waterways	16.420	0.027	−7%	0.0016	0.0016	10	15%
Railways	22.257	0.039	3%	0.0012	0.0012	55	20%
Total ton-km	−1%						

	Welfare components (% change)[a]		
Utility	0.11%	Tax income passenger transport	40%
Pollution passenger transport	−0.10%	Tax income freight transport	59%
Pollution freight transport	−1.65%		
Accident costs	−2.52%		
Road damage	−3.46%		
Welfare gain	0.87 million ECU/day		

[a] % change with respect to reference values.

we investigated the effectiveness of public transport pricing. Finally, we determined the full optimum, in which all price instruments are available, and, in addition, the possibility of introducing specific new technologies are optimally selected.

2.4.1. *Pricing optimum*

We first consider a 'pricing' optimum. We solve the model to determine the set of prices that maximizes the objective function without considering the choice of technologies, where it is assumed that each mode and period can be priced differently. The pricing optimum so obtained is given in Table 2.4. The results are easily summarized. First, all prices in the peak period for both passenger and freight transport in the peak have risen compared with the reference situation. Moreover, it is proportionately larger for small cars, resulting in a smaller price difference between small and large cars. Also note that the price differential between petrol and diesel decreases relative to the reference situation. These findings can be explained by the fact that, especially in the peak period, the main external cost of cars is the contribution to congestion and that, in the model, this contribution is assumed to be exactly the same for different types of cars. Moreover, differences in pollution costs are quite small, and accident costs were assumed to be the same for all car types (Mayeres et al., 1996). While differences in social external costs cannot justify large tax

Table 2.5. No toll system available.

		Pass-km/ day (million)	Price (ECU/ pass-km)	% change[a] (ECU/ pass-km)	Tax (ECU/ pass-km)	Marginal external cost	Speed (km/h)	Market share
Passenger transport								
Private transport								
Peak								
Large	Petrol	18.305	0.379	7%	0.143	0.220		
	Diesel	10.557	0.334	18%	0.134	0.217	56.8	42%
Small	Petrol	27.416	0.239	11%	0.095	0.217		
	Diesel	16.015	0.212	17%	0.086	0.213		
	Total	−2%						
Off-peak								
Large	Petrol	18.761	0.385	8%	0.148	0.063		
	Diesel	10.522	0.342	20%	0.141	0.059	95.7	42%
Small	Petrol	27.540	0.250	12%	0.104	0.061		
	Diesel	15.512	0.225	21%	0.097	0.056		
	Total	−5%						
Technology								
Large	Petrol				Standard			
	Diesel				Standard			
Small	Petrol				Standard			
	Diesel				Standard			
Public transport								
Peak								
Bus		6.059	0.039	32%	−0.0021	0.0120	44	4%
Train		6.236	0.078	68%	−0.0285	0.0003	70	4%
Total		−12%						
Off-peak								
Bus		6.820	0.039	32%	0.0125	0.0054	73.7	4%
Train		7.724	0.078	68%	0.0231	0.0007	70.0	5%
Total		−13%						
Total pass-km		−5%						

Continued

differences between large and small cars, this result may change once distributional issues are taken into account.

In the off-peak period, several car prices actually decline as a consequence of the fact that in the 1991 reference situation they paid more in taxes than the external cost they imposed. The price increases for public transport of both passengers and freight can be explained by the fact that subsidies in the reference situation implied that users did not even pay the private costs. On the contrary, the price of inland waterway transport would decline slightly. Finally note that optimal pricing would have a non-negligible impact on transport flows. Peak car transport would decline by some 11%, off-peak car

Table 2.5. (*Continued.*)

Freight transport							
	Ton-km/ day (million)	Price (ECU/ ton-km)	% change[a] (ECU/ ton-km)	Tax (ECU/ ton-km)	Marginal external cost	Speed (km/h)	Market share
Road	13.802	0.063	9%	0.0127	0.0378	37	65%
Peak							
Off-peak	58.182	0.063	9%	0.0127	0.0093	62	
Waterways	16.660	0.027	−8%	0.0013	0.0016	10	15%
Railways	22.583	0.039	2%	0.0008	0.0012	55	20%
Total ton-km	−1%						

Welfare components (% change)[a]			
Utility	−0.01%	Tax income passenger transport	44%
Pollution passenger transport	−3.27%	Tax income freight transport	52%
Pollution freight transport	−1.89%		
Accident costs	−3.91%		
Road damage	−4.44%		
Welfare gain	0.20 million ECU/day		

[a] % change with respect to reference values.

traffic would increase by 7%. Moreover, public transport use, and especially peak truck traffic, would both go down.

Table 2.4 also summarizes the impact of the optimal prices on consumer utility, on various external effects, and on tax revenues. All external costs of the transport sector are lower in the optimum than in the reference situation, although the relatively small reduction in overall traffic flows implies modest reductions in pollution. Tax revenues rise substantially, especially those generated on freight flows. Importantly, the results indicate that the implementation of the optimal pricing policies would yield a welfare gain of approximately 0.87 million ECU per day.

2.4.2. *Optimal pricing when no toll is available*

It is possible that pricing instruments that allow optimal differentiation between peak and off-peak prices (tolls, road pricing) are not implementable. In this subsection we assume that no such toll is available. Moreover, we also assumed that it is not feasible to charge different public transport prices in peak and off-peak periods. The resulting optimum is described in Table 2.5. Although the general trends are similar to those of the previous case, there are some obvious differences. The absence of a toll no longer allows the fine-tuning of the tax system according to temporal variations in externalities. The remaining

Table 2.6. Optimal choice of technology.

		Passenger transport						
		Pass-km/ day (million)	Price (ECU/ pass-km)	% change[a] (ECU/ pass-km)	Tax (ECU/ pass-km)	Marginal external cost	Speed (km/h)	Market share
Private transport								
Peak								
Large	Petrol	18.095	0.353	0%	0.108	0.224		
	Diesel	11.563	0.282	0%	0.082	0.231	55.1	41%
Small	Petrol	27.491	0.216	0%	0.063	0.223		
	Diesel	16.857	0.180	0%	0.054	0.227		
	Total	−0.07%						
Off-peak								
Large	Petrol	18.621	0.357	0%	0.111	0.052		
	Diesel	11.898	0.285	0%	0.084	0.060	95.0	42%
Small	Petrol	28.287	0.223	0%	0.068	0.052		
	Diesel	17.344	0.186	0%	0.058	0.057		
	Total	−0.10%						
Technology								
Large	Petrol				Improved			
	Diesel				Standard			
Small	Petrol				Improved			
	Diesel				Standard			
Public transport								
Peak								
Bus		5.461	0.030	0%	−0.012	0.013	42	3%
Train		8.491	0.046	0%	−0.060	0.000	70	5%
Total		−0.08%						
Off-peak								
Bus		6.497	0.030	0%	0.003	0.005	73.2	4%
Train		10.112	0.046	0%	−0.008	0.001	70.0	6%
Total		−0.10%						
Total pass-km		−0.09%						

Continued

instruments serve as very imperfect tools to correct for both congestion and environmental damage. Note that all prices rise compared with the reference situation, but differences between periods vanish. Prices exceed marginal social costs in the off-peak period, and travellers in the peak pay a tax which is much smaller than the corresponding marginal external cost. Moreover, Table 2.5 suggests that the additional restriction on available tax instruments reduced the welfare gain from 0.87 to 0.20 million ECU per day. In other words, the absence of a toll reduces the potential welfare gain dramatically.

Table 2.6. (*Continued.*)

Freight transport							
	Ton-km/ day (million)	Price (ECU/ ton-km)	% change (ECU/ ton-km)	Tax (ECU/ ton-km)	Marginal external cost	Speed (km/h)	Market share
Road							
Peak	14.022	0.058	0%	0.008	0.040	36	67%
Off-peak	61.245	0.058	0%	0.008	0.010	62	
Waterways	14.291	0.029	0%	0.004	0.002	10	13%
Railways	22.313	0.038	0%	0.000	0.001	55	20%
Total ton-km	−0.09%						

Welfare components (% change)[a]			
Utility	−0.09%	Tax income passenger transport	−7%
Pollution passenger transport	−49.26%	Tax income freight transport	0%
Pollution freight transport	−0.09%		
Accident costs	−0.09%		
Road damage	−0.08%		

[a] % change with respect to reference values

2.4.3. *Introducing new technologies*

In a third exercise, we introduced the possibility of allowing the government to implement specific new technologies. For illustrative purposes we only consider a very simple example. Regarding petrol-fuelled cars the desirability of subsidizing the introduction of catalytic converters was evaluated.[14] For diesel cars we considered subsidizing an improved engine technology (turbo-cooler, installation of an oxidation cat). These cleaner technologies imply higher resource costs, but they are less pollutant. Should the new technology be subsidized, prices would remain unchanged, but tax revenue and pollution costs would be different.

The results (Table 2.6) suggest that, evaluated at 1991 figures, it would be useful to subsidize installation of catalytic converters on all petrol-fuelled cars. The reduction in pollution obtained by this policy would more than compensate for the reduction in overall tax revenues associated with the subsidies. However, the reduction in pollution due to improved diesel technology does not outweigh the higher resource cost. Note that prices and taxes are the same as in the reference situation. The results further suggest that the policy being considered would substantially reduce pollution associated with passenger transport at almost constant traffic flows. However, this would imply a welfare gain of only 0.11 million ECUs per day, which amounts to only 13% of the welfare gain at the full pricing optimum.

Table 2.7. Optimal public transport prices.

		Pass-km/ day (million)	Price (ECU/ pass-km)	% change[a] (ECU/ pass-km)	Tax (ECU/ pass-km)	Marginal external cost	Speed (km/h)	Market share
Passenger transport								
Private transport								
Peak								
Large	Petrol	18.167	0.353	0%	0.117	0.237		
	Diesel	11.605	0.282	0%	0.082	0.233	54.8	42%
Small	Petrol	27.581	0.216	0%	0.072	0.234		
	Diesel	16.906	0.180	0%	0.054	0.230		
	Total	0%						
Off-peak								
Large	Petrol	18.708	0.357	0%	0.120	0.064		
	Diesel	11.954	0.285	0%	0.084	0.060	95.0	43%
Small	Petrol	28.418	0.223	0%	0.077	0.062		
	Diesel	17.423	0.186	0%	0.058	0.057		
	Total	0%						
Technology								
Large	Petrol				Standard			
	Diesel				Standard			
Small	Petrol				Standard			
	Diesel				Standard			
Public transport								
Peak								
Bus		5.765	0.046	55%	0.005	0.013	42	3%
Train		5.135	0.099	114%	−0.007	0.000	70	3%
Total		−22%						
Off-peak								
Bus		6.861	0.028	−3%	0.002	0.005	73.1	4%
Train		9.604	0.051	9%	−0.004	0.001	70.0	5%
Total		−1%						
Total pass-km		−2%						

Continued

2.4.4. *Optimal public transport pricing*

In this exercise, we let the model optimally determine public transport (train and bus) prices, keeping all other prices unchanged at their reference level. Results are given in Table 2.7. Public transport prices rise, but by less than in the pricing optimum. There remain subsidies for passenger transport by train, but they are less important than in the reference situation. This is a classic second-best result: as other modes are underpriced (as compared to the optimum), it is optimal not to make public transport users pay the the full marginale social cost. Table 2.7 reveals, however, that using only public transport prices

Table 2.7. (Continued.)

Freight transport							
	Ton-km/ day (million)	Price (ECU/ ton-km)	% change[a] (ECU/ ton-km)	Tax (ECU/ ton-km)	Marginal external cost	Speed (km/h)	Market share
Road	13.998	0.058	0%	0.008	0.041	36	67%
Peak							
Off-peak	61.335	0.058	0%	0.008	0.010	62	
Waterways	14.323	0.029	0%	0.004	0.002	10	13%
Railways	22.362	0.038	0%	0.000	0.001	55	20%
Total ton-km	0%						

Welfare components (% change)[a]			
Utility	0.02%	Tax income passenger transport	7%
Pollution passenger transport	0.31%	Tax income freight transport	0%
Pollution freight transport	0.02%		
Accident costs	0.33%		
Road damage	0.01%		
Welfare gain	0.11 million ECU/day		

[a] % change with respect to reference values.

is not an efficient policy: it attains only 13% of the welfare gain that is attainable in the full pricing optimum.

2.4.5. *Full optimum: optimal pricing and choice of technology*

We finally consider the case where the government has all pricing instruments available and, in addition, it can freely decide whether or not to subsidize catalytic converters and improved diesel technologies. The results are given in Table 2.8. Not surprisingly, they combine the results found in sections 2.4.1 and 2.4.3. Taxes are identical to those discussed in section 2.4.1, except for some minor differences related to big petrol-fuelled cars, and catalytic converters are introduced for all petrol-fuelled cars. The welfare gain amounts to 0.99 million ECU per day.

2.5. Conclusions

In this paper we looked for optimal pricing and regulatory policies in the transport sector within the framework of a standard welfare maximization problem. First, a simple theoretical model was developed. We then constructed a simulation model consistent with the theoretical framework to study a variety of policies using data on Belgian inter-regional transport. The model allowed

Table 2.8. Full optimum.

Passenger transport								
		Pass-km/ day (million)	Price (ECU/ pass-km)	% change[a] (ECU/ pass-km)	Tax (ECU/ pass-km)	Marginal external cost	Speed (km/h)	Market share
Private transport								
Peak								
Large	Petrol	18.714	0.405	15%	0.161	0.161		
	Diesel	10.506	0.367	30%	0.168	0.168	63.7	38%
Small	Petrol	23.250	0.313	45%	0.160	0.160		
	Diesel	13.099	0.290	61%	0.164	0.164		
	Total	−11%						
Off-peak								
Large	Petrol	22.388	0.300	−16%	0.054	0.054		
	Diesel	12.685	0.262	−8%	0.061	0.061	94.2	47%
Small	Petrol	29.633	0.209	−6%	0.054	0.053		
	Diesel	16.678	0.187	1%	0.059	0.059		
	Total	7%						
Technology								
Large	Petrol	Improved						
	Diesel	Standard						
Small	Petrol	Improved						
	Diesel	Standard						
Public transport								
Peak								
Bus		6.404	0.051	71%	0.0094	0.0094	49	4%
Train		4.787	0.107	130%	0.0003	0.0003	70	3%
Total		−20%						
Off-peak								
Bus		6.666	0.032	9%	0.0056	0.0056	72.6	4%
Train		9.408	0.055	19%	0.0007	0.0007	70.0	5%
Total		−3%						
Total pass-km		−4%						

Continued

for a large number of transport alternatives through the use of nested CES utility and cost functions, and it captured the most important external effects associated with transport services. The model was used to determine optimal taxes and public transport prices, optimal choice of technologies, and combinations of the two policies.

An advantage of the type of model developed here is that it directly provides a monetary measure of overall welfare. As a consequence, the model could be used to compare the welfare implications of, for example, implementing a toll system (pricing policy), priority bus lanes (traffic management), investment in

Table 2.8 (*Continued.*)

Freight transport							
	Ton-km/ day (million)	Price (ECU/ ton-km)	% change[a] (ECU/ ton-km)	Tax (ECU/ ton-km)	Marginal external cost	Speed (km/h)	Market share
Road							
Peak	12.257	0.080	37%	0.0292	0.0292	41	65%
Off-peak	60.392	0.061	4%	0.0098	0.0098	61	
Waterways	16.403	0.027	−7%	0.0016	0.0016	10	15%
Railways	22.233	0.039	3%	0.0012	0.0012	55	20%
Total ton-km	−1%						

Welfare components (% change)[a]			
Utility	0.01%	Tax income passenger transport	31%
Pollution passenger transport	−50.75%	Tax income freight transport	59%
Pollution freight transport	−1.75%		
Accident costs	−2.50%		
Road damage	−3.56%		
Welfare gain	0.99 million ECU/day		

[a] % change with respect to reference values.

inland waterways capacity (infrastructure policies), imposing environmental standards and introducing new technologies.

Notes

1. This paper presents results of the TRENEN-project financed by the JOULE II program of the European Community. The project is coordinated by Stef Proost (KULeuven). We are grateful to S. Ochelen, R. Roson, S. Proost for useful discussions, and to two anonymous referees for detailed written comments.
2. Allowing for decreasing returns would imply the possibility of positive private sector profits. This substantially complicates the theoretical analysis, because account has to be taken of the distributional implications of profits (see, e.g. Yang, 1993; De Borger, 1997).
3. The constancy of private and external costs other than congestion is of course not necessary from a theoretical perspective. We made this assumption because data limitations forced us to impose it in the empirical application.
4. In practice, of course, some transport services (e.g. rail transport) do not contribute to congestion. In that case, just think of the partial derivative of g with respect to the use of this particular mode being equal to zero.
5. Another possibility would be to maximize welfare subject to a budgetary constraint. The advantage of directly incorporating the budgetary implications of transport and environmental policies into the objective function is that the level of tax income is endogenously optimized. The disadvantage is that the choice of the shadow cost to be applied is difficult to determine *a priori* as it depends on what tax instruments the authorities have available.
6. These assumptions are commonly used in the literature to ease the interpretation. Note,

however, that especially with respect to freight transport they are quite unrealistic. Even if conditional demands are insensitive to prices of other freight transport services, a freight rate change implies an induced output effect as well.

7. Spatial differences in social costs and transport flows were taken into account within the TRENEN project by developing two companion models: the inter-regional model discussed here, and the urban model. The inter-regional transportation market is characterized by lower market shares for public transport, less congestion and noise costs, and a much greater importance of freight transport (see De Borger et al., 1995).
8. A number of sensitivity analyses were carried out to see how sensitive the results were with respect to the assumed elasticity values. We found that the optimal transport volumes were quite sensitive to differences in elasticities. However, the optimal prices were found to be much less affected, except for extremely large changes in own price elasticities. Results are available from the authors.
9. The values of time used are taken from from Hague Consulting Group (1992).
10. A car is equal to one PCU, while buses and trucks equal two PCUs.
11. Note that in order to close the model at the calibration stage a third (untaxed) consumer good was introduced in the utility function apart from passenger transport and the aggregate private good produced by the private production sector. Although this commodity has no impact on the optimal pricing results of the simulations it may be interpreted as a good the production of which does not generate any demand for freight transport (say, services).
12. In order to make the presentation and the interpretation of different transport outcomes more transparent, not all transport markets have been incorporated in the tables that will be presented.
13. These three instruments suffice to implement the full pricing optimum. See De Borger and Swysen (1996) for details. Note that we use the toll as a possible instrument that allows price diferentiation between peak and off-peak periods.
14. TRENEN can include a large number of different car technology options. The major problem up to now was not the modelling but the cost data for alternative emission technologies.

References

De Borger, B. and D. Swysen, 1996, TRENEN-Interregional model documentation. SESO-UFSIA, Antwerpen, January 1996.

De Borger, B., 1997, Pricing of Public Final and Intermediate Goods in the presence of externalities, European Journal of Political Economy, forthcoming.

De Borger, B., I. Mayeres, S. Proost and S. Wouters, 1996, Optimal Pricing of Urban Passenger Transport: a simulation exercise for Belgium, Journal of Transport Economics and Policy, XXX, 31–53.

De Borger, B., S. Ochelen, S. Proost and D. Swysen, 1995, Optimal Policies for Curbing Congestion and Air pollution in Urban Transport, Working Paper. Leuven: CES-KU.

Bovenberg, L. and R. Van der Ploeg, 1994, Environmental policy, public finance and the labour market in a second best world', Journal of Public Economics, 55, 349–390.

Goodwin, P.B., 1992, A review of new demand elasticities with special reference to short and long run effects of price changes, Journal of Transport Economics and Policy, 26, 155–169.

Hague Consulting Group, 1992, De reistijdwaardering in het goederenvervoer, Den Haag, Rapport Hoofdonderzoek, HCG.

Keller, W.J., 1976, A nested CES-type utility function and its demand and price index functions, European Economic Review, 7, 75–186.

Kirwan, K.J., M.M. O'Mahony and D. O'Sullivan, 1995, Speed-Flow Relationships for Use in an Urban Transport Policy Assessment Model, Working paper. Dublin: Trinity College.

Mayeres, I., 1993, The marginal external costs of car use – with an application to Belgium, Tijdschrift voor Economie en Management, XXXVIII, 225–258.

Mayeres, I. and S. Proost, 1996, Optimal tax and tax reform rules with congestion type of externalities, Leuven: KUL-CES.

Mayeres, I., S. Ochelen and S. Proost, 1996, The marginal external costs of transport revisited, Transportation Research (series D), forthcoming.

Newbery, D.M.G., 1988, Road User Charges in Britain, The Economic Journal, 98, 161–176.

Oum, T.H. and M.W. Tretheway, 1988, Ramsey Pricing in the presence of externality costs, Journal of Transport Economics and Policy, 22, 307–317.

Oum, T.H, W.G. Waters and J.S. Yong, 1992, Concepts of price elasticities of transport demand and recent empirical estimates, Journal of Transport Economics and Policy 26, 139–154.

Sandmo, A., 1975, Optimal taxation in the presence of externalities, Swedish Journal of Economics, 77, 86–98.

Small, K.A. and C. Kazimi, 1995, On the costs of air pollution from motor vehicles, Journal of Transport Economics and Policy, 29, 7–32.

Wijkander, H., 1985, Correcting externalities through taxes on/subsidies to related goods, Journal of Public Economics, 28, 111–125.

Yang, C.C., 1993, Distributional equity and the pricing of public final and intermediate goods, Economic Letters, 41, 429–434.

CHAPTER 3

Revealed Preferences, Externalities and Optimal Pricing for Urban Transportation[1]

Roberto Roson

3.1. Introduction

What is the right price for transportation services? Economic theory suggests that prices are correctly set and resources are efficiently allocated when prices reflect social marginal costs. If market prices are distorted, from a social point of view, 'Pigouvian' taxes and subsidies could be introduced to bring prices in line with social marginal costs. In principle, the determination of such an optimal tax structure would imply the identification of all positive and negative externalities, and the valuation of all of them. In practice, things are more complicated. For urban transportation services, for example, difficulties may arise because (1) it may not be easy to single out all positive and negative externalities associated with transport activities, especially in the case of negative environmental externalities, because the impact on the environment of some pollutants may be uncertain; (2) the valuation of externalities may be subjective, because of the need to attach values to things like human health and life. Alternative methods can then be used to assess the monetary value of externalities, but there is not a single best method available for all circumstances; (3) marginal costs may not be constant. If marginal costs depend on activity levels, demand–supply interactions in the markets should be taken into account. This is made more complicated by the fact that the market for transport services is a very complex one. Transport modes are imperfect substitutes and are sometimes linked by complementarity relationships. Private transport services co-exist with publicly supplied services, whose prices are not determined by the market. Natural and local monopolies can be found alongside highly competitive sub-markets, and political and technical constraints also influence market mechanisms.

In this paper we use an applied optimization model, named TRENEN, especially designed to identify Pigouvian taxes in the presence of non-linear

R. Roson and K.A. Small (eds.), Environment and Transport in Economic Modelling, 39–60.

marginal costs and of various policy constraints. Optimal prices are computed via welfare optimization, by considering all different resource and external costs associated with a set of alternative transport modes. Equilibrium prices and marginal costs are estimated within a multiple market partial equilibrium framework. The urban version of this model model has been implemented for the cities of Brussels, Dublin, Amsterdam, Mestre and Bologna. The Bologna implementation is considered here.

Our first simulation results suggest that market prices in the calibration year of the model (1991), inclusive of taxes and subsidies, are very distorted. Correspondingly, when the model computes an optimal price structure for transport services, it implicitly carries out two distinct optimization stages: (1) a realignment of current prices to resource costs and revealed preferences, and (2) the internalization of external costs. We found that the first stage is often the most important one in explaining potential welfare gains.

One reason for market distortions is the existence of external policy objectives, affecting fiscal policies in the transport sector. For example, fuel taxes may be influenced by international trade policies, and indirect taxation of transport services may be affected by income distributional policies. Consistency and compatibility between external policy objectives, market-specific policies and protection of the environment shoud be carefully assessed.

In this paper, we use the TRENEN model to analyse the effect of market distortions and external policy objectives on the process of internalization of external costs of urban transport activities. We shall focus, in particular, on implicit distributional policies associated with indirect taxation.

3.2. Urban transport services in Italy: institutions and market distortions

Prices in the market for transport services are affected by a variety of taxes and subsidies. These taxes and subsidies have been introduced for different purposes, at different times, and by different institutions. In order to better understand the model results, which refer to an Italian city, we provide here a brief account of the main institutional mechanisms underlying the existence of market-distorting fiscal policies in Italy.

According to the Italian Constitution, local regional governments have the right to issue laws on some specific topics, including transportation. In practice, given the limited geographical extension of each region, this right has been applied only to urban and metropolitan transport.[2] Regional governments were established in the late 1970s and started to implement transport policies in the early 1980s, also co-ordinating the action of sub-regional bodies like provinces and municipalities. At that time the state assigned quite large subsidies, in different ways, to the private and public suppliers of transport services. In order to rationalize access to public funds, a single national fund for financing deficits of public transport companies was established and distributed to the regions according to the total volume of services supplied.

In principle, each company could benefit from a state contribution equal to 65% of its standard costs, obtained by multiplying a standard unit cost by the total distance supplied (in km). In the distribution scheme no reference was made to the actual number of carried passengers, to the quality of services, to the geographical coverage, or to the type of roads and surfaces (e.g. hills or plain). In addition, most regions gave preferential access to the subsidies to public-owned companies, and encouraged the merging of small companies and the creation of consortia, which could be better controlled by local governments. Many private companies were sold to public bodies.

The amount of subsidies fixed at about 65% of operating costs was not determined on the basis of some kind of economic analysis, but on the basis of the existing situation. The very generous support of the state, nonetheless, proved to be largely insufficient during the 1980s, because of a decline in the demand for public transport, an increase in the supply (total km) of the transport companies, and increasing labour costs, especially after the collective labour contract signed by the trade unions in October 1989 (Cobello, 1994).

In order to avoid the bankruptcy of a large number of local transport companies, the state assigned additional funds in 1987 and 1991. By doing this, the state acted as a 'lender of last resort', financing large deficits created by other public bodies. In other words, regions and other local governments were allowed to 'free-ride', resulting in inefficient management of local public transport without bearing full financial responsibility.

This lack of co-ordination also applies to railway transport, which in some metropolitan areas (like Bologna) can be considered part of the local transport system. Railway transport is directly controlled by the state via the Ministry of Transportation, but within the ministry, an autonomous body was established for the management of the national railway network (Azienda Autonoma Ferrovie dello Stato).

This body was transformed to Ente F.S. in 1985. This new body was still part of the central administration of the state, but it had a larger degree of autonomy and, more importantly, was supposed to follow management criteria similar to those of a private company. This transformation phase was mainly induced by financial problems. Whereas the selling of transportation services accounted for about 51.6% of total operating costs in 1972, in 1980 market revenues covered only 31.9% of costs, falling to 27.3% in 1990 (Ministero dei Trasporti, 1994). This rapid deterioration of the cost/revenue balance was partly due to the reasons discussed above for local transport companies: declining demand, poor quality of services, increasing costs (especially labour costs). In addition, railway fares were much more difficult to update during periods of high inflation (mainly for political reasons).

In 1992, a second major transformation occurred. The managing body was transformed in a private corporation, with 100% of shares still owned by the

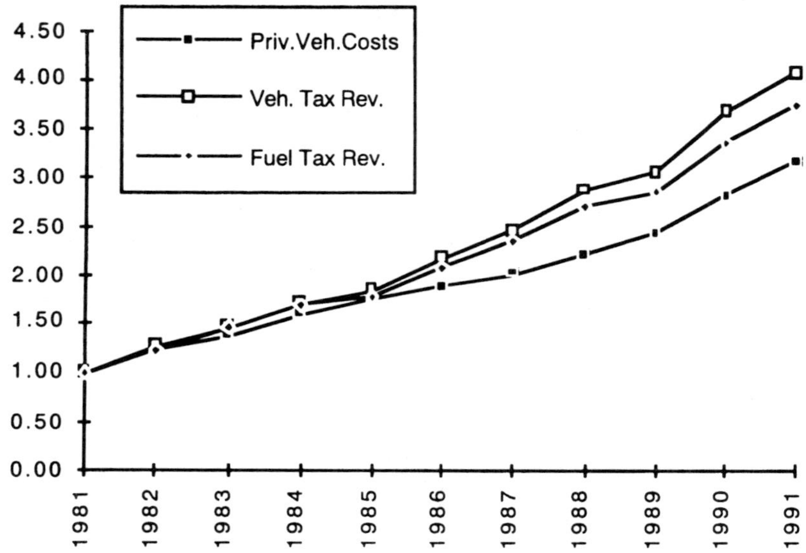

Figure 3.1. Trends in private car taxation and costs.

state. Following the European directive 440/91, the network infrastructure management has been separated from the transport service management, and the transfer of funds from the state and regions has been regulated by a special contract. This contract specifies which kinds of 'social services' will be supplied (including reduced fares) and what the public bodies will pay for them.

What we can see now is a difficult attempt of rationalization of the public transport sector. Institutional mechanisms and political pressure groups have both contributed to create, especially during the 1980s, a very unbalanced situation, in which public funding seems to have fallen out of control.

In contrast, taxes on private transport are traditionally higher in Italy than in other European countries, and have increased further during the last decade. An annual tax is paid by car owners, the amount depending on the engine power, and another annual tax is paid by each driver on his driving licence. The most important tax is, however, the set of taxes and excises on fuels.

Taxes on private transport were substantially raised during the 1970s, after the first oil crisis. In 1981 taxation on vehicles generated a revenue of 14 486 billion lire, of which 69% was due to fuel taxes, while total operating costs of private vehicles were 32 432 billion. In 1991, tax revenue was 58 702 billion (+305% at current prices, where fuel taxes account for 63%) and operating costs of private vehicles amounted to 102 853 billion (+218%; source: Conto Nazionale dei Trasporti). Taxation on vehicles, then, increased faster than total operating costs in the period 1981–1991 (Figure 3.1). Figure 3.1 also shows the upward jumps in the taxation of private vehicles which occurred in 1986 and 1990. Remarkably, over the same periods the state was forced to intervene to avoid the collapse of the public transport system. High and increasing taxation on private vehicles seems, therefore, to mirror the increasing high level of subsidy of public transport.

From this perspective, fiscal policies appear to have acted as a major mechanism of market distortion. This is probably because the political debate has focused on 'if' and 'how' a public intervention in the transport sector would be beneficial, and it has never tackled issues of 'how much'. Positive externalities for public transport and negative externalities for private transport have often been pointed out as a justification for the support of a subsidized public transport system, but these externalities have never been quantified.

3. The TRENEN–urban model

TRENEN (**tr**ansportation, **en**ergy and the **en**vironment) is an optimization model for the assessment of fiscal policies in the transport sector. Developed in two versions (urban and inter-regional) within the European research project JOULE II, it is currently under further development in the context of the transport research project of the IV Framework Programme. The basic version of the TRENEN–urban model has been implemented for various European cities, including the Italian city of Bologna.

The structure of TRENEN–inter-regional is described in detail in Chapter 2. The urban version of the model is similar.[3] The modelling of consumer demand is basically the same, although different transport modes are considered in the two cases and the structure of demand is different.[4] In addition, the inter-regional model contains a sub-model in which the demand for freight transportation is made endogenous, which is not included in the urban model.[5]

The TRENEN model adopts a welfare maximization approach to determine optimal taxation levels for a set of alternative transport modes. The model is based on the determination of social welfare as a difference between money-metric consumer utility (including redistributed tax revenue) and total production costs (including externalities). Demand and supply side of the model are briefly described here below.

3.3.1. *Demand side*

One or more representative consumers are considered in the model. Each consumer makes a choice between a set of goods and services, which are regarded as imperfect substitutes. These goods include a composite good, accounting for all non-trasport consumption, and several urban transportation services, specified by mode and differentiated by period of the day (peak/off-peak) and other characteristics (e.g. number of passengers, existence of a catalytic converter, vehicle size, fuel type).

The modelling of the consumer choice is hierarchical: first, available income

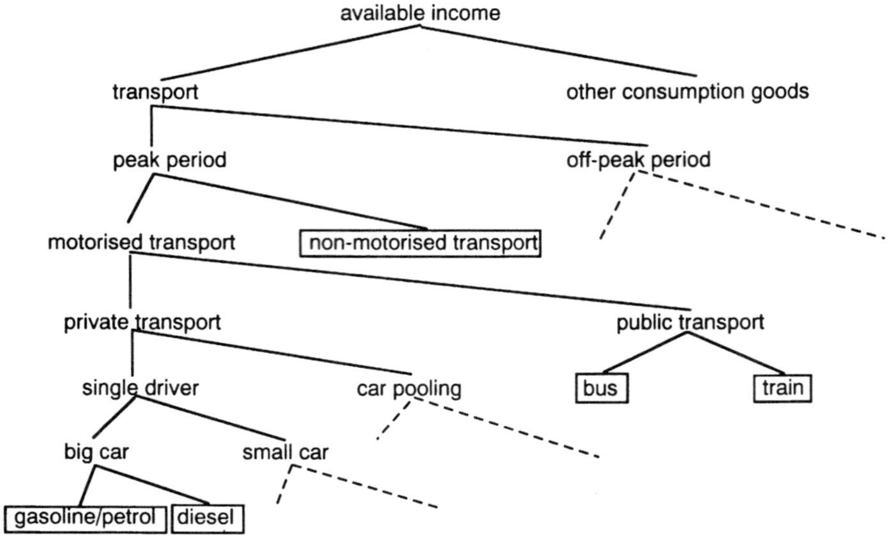

Figure 3.2. The nested demand structure of TRENEN–urban.

is allocated between consumption of transport and other goods; second, a choice is made between travel periods, then the option between private and public transport is considered; third, alternative private and public modes are chosen, and so on. Each stage of the decision process is modelled by mean of a CES function. As a consequence, consumption shares are always positive but sensitive to (generalized) prices, depending on given price elasticity parameters.

A CES utility (production) function expresses utility (output, Q) as a weighted average of consumption levels (inputs, q):

$$Q = \left(\sum_{i=1}^{n} \omega_i q_i^{\rho} \right)^{1/\rho}$$

The level of consumer utility is computed through a bottom-up process (see Figure 3.2). Starting from generalized money prices (including time costs), the model computes demand shares for the bottom levels of the decision tree. On the basis of demand shares and prices, price and quantity indexes are computed for the upper decision level and the choice process is repeated recursively (Keller, 1976). Each price index is obtained through a cost/expenditure function, expressing the minimum cost which is necessary to reach a given utility level.[6]

At the top level, the simulated price index P_s includes the generalized prices of all modes. The aggregate consumption quantity is then measured as exogenous available income, Y_0, divided by P_s; when multiplied by the initial price index P_0 it provides an 'equivalent income' index $P_0 Y_0 / P_s$. This indicates what income level would have been considered equivalent, by the consumer, to real income under the new set of market prices, if prices had not been altered. This can be taken as a monetary measure of utility, allowing one to express total welfare as a difference between direct monetary utility and external costs. Finally, changes in tax revenue ΔT are assumed redistributed to consumers and therefore are added to their money income. This effectively combines changes in consumer surplus with tax revenue.

3.3.2. *Supply side*

Generalized prices for all transport modes considered in the model are computed by adding up different cost elements. For example, generalized prices for private vehicles include average purchase, maintenance and other operating costs, parking costs, time costs and taxes.[7] Average time costs per period of the day are estimated via a macro-congestion function of the following type:

$$t = \alpha + \beta\, e^{\gamma \mathrm{PCU}} \tag{3.1}$$

where $t =$ time needed to drive 1 km, PCU are 'passenger car units' (equivalent car units; a measure of traffic volume), and all remaining symbols are estimated parameters.[8]

Optimal tax levels are determined within the model. In the unconstrained version, it is assumed that is is possible to discriminate between two periods of the day (peak and off-peak), thereby simulating a simple road pricing scheme, as well as on the basis of the number of passengers (single driver, car pooling) for private vehicles.[9]

3.3.3. *Equilibrium*

Optimal tax levels for the various transport modes are computed by maximizing a measure of social welfare, including consumer equivalent income, accident and pollution external costs. This welfare index is used to compute the market equilibrium point, where total net surplus is maximized. Each vehicle type is associated with a set of emissions/km, and pollution costs are computed by multiplying emission levels by the social marginal cost of each pollutant. Social welfare is defined as:

$$W_s = P_0 \frac{Y_0 + \Delta T_s}{P_s} - E_s \tag{3.2}$$

where $E =$ total external costs (accident and pollution, whereas congestion is accounted for in the generalized prices). The first part of the welfare function includes a money-metric measure of the utility for the representative consumer, expressed as equivalent income. The second part is a vector of external costs, in monetary terms. Because of the existence of a single agent in the economy, extra profits and producer surplus are ruled out (in other words, all market prices are determined on the basis of marginal production costs).

The model can simulate the adoption of new, cleaner technologies. Such technologies often imply higher private costs but lower external costs. Internalization taxes could then make the adoption of cleaner technologies convenient. In the current model version, the option of introducing catalytic converters on some cars is taken into account. This is done by adding an extra layer at the bottom of the decision tree (Figure 3.2), expressing the option of introducing catalytic converters. In this case, values for the elasticity parameters are set very high, so the demand goes all to the cheaper alternative. Issues of dynamic adjustment and car stock replacement are not explicitly considered.

3.3.4. *Calibration*

To understand how reference prices and quantities influence the simulation results, it is important to recall that most of the structural parameters in TRENEN are estimated using a calibration procedure. This amounts to selecting parameter values that make the model consistent with the observed behaviour in a certain period of time (a year).[10] For example, from an observed price–quantity pair, a calibration procedure sets demand parameters in such a way that the observed point can be found along a given demand function.

A different but equivalent interpretation of the calibration method is in terms of utility and revealed preferences. Since consumption demand stems from a utility maximization problem, the identification of a demand function implies the estimation of the marginal utility contribution of each consumed good or service. In other words, the calibration method used in TRENEN allows us to assess the revealed utility value of each transport service, as it stems from actual consumer behaviour.

Consider, for example, the hierarchical decision process which is used to model the consumption of the different transport services. At each level of the decision tree, a choice is made about the relative demand of two or more consumption alternatives. Given the CES functional form adopted, the relative level of demand is:

$$\frac{q_i}{q_j} = \left(\frac{\omega_i}{\omega_j}\frac{p_j}{p_i}\right)^{\sigma} \tag{3.3}$$

where q = demand quantities, p = prices, and all remaining symbols are parameters.

Three parameters are involved in the equations: an elasticity parameter $\sigma = 1/1 - \rho$, and two 'weight' parameters ω. From standard consumer theory it is possible to show that weight parameters can be normalized, so that only one of the two weight parameters must actually be estimated. The elasticity parameter expresses the degree of product differentiation (goods are regarded as imperfect substitutes), as well as the substitution possibilities offered by the transportation system.[11]

At a certain calibration year, one can observe total consumption volumes and prices, possibly including taxes or subsidies. So there is only one ratio of weight parameters for which equation (3.3) is satisfied in the base year, for a chosen value of the elasticity parameter. The weight parameters express the marginal utility associated with the consumption of each good. For example, if the price for a good in equation (3.3) is relatively low, possibly because of a subsidy, but the demanded quantitity is also relatively low, then a small weight parameter is associated to that good. The calibration procedure therefore infers consumer preferences from the actual consumption behaviour. This has some important implications for the simulation experiments described here, because market shares of alternative transport modes in the calibration year of the model do not appear to be significantly infuenced by relative prices.

3.4. Results

3.4.1. The basic unconstrained simulation experiment

A version of the TRENEN–urban model implemented for Bologna, with a single representative consumer, has been used to run a benchmark simulation in which optimal taxation for urban transportation services is determined. The optimization programme does not include, in this case, any technical or political constraint, so a first-best solution with social marginal cost pricing is identified (Tables 3.1 and 3.2).

We found that, when optimal prices are introduced, total pollution costs are reduced by 27%,[12] whereas total accident costs are reduced by 6%. An index of generalized prices for urban transport services increases by 0.67%[13] but, remarkably, the overall consumer equivalent income also increases. This phenomenum is due to the elimination of market distortions and to the redistribution of tax revenue, which allows a better allocation of consumption expenditure. This positive effect outweights the reduction of real income due to the introduction of externality taxes.

The urban transport sector as a whole is no longer subsidized in the optimum state. In particular, subsidies assigned to buses and trains are completely eliminated, and public transport is actually slightly taxed. This result is consistent with the hypothesis of low, but positive external costs of public transport.[14] Private transport services continue to be taxed, but total tax revenue obtained from this source is reduced by 17%.

Changes in demand levels are symmetrical to changes in relative generalized prices. The demand for bus services declines by 19% in the peak period and by 50% in the off-peak, whereas demand for train services is reduced by 28% and 58%, respectively.

The composition of private vehicle traffic is also changed in the new equilibrium. Private vehicles are used relatively more during the off-peak period, when more passengers are carried. Demand for diesel cars increases for all periods and all car types, mainly because of the elimination of the tax discrimination

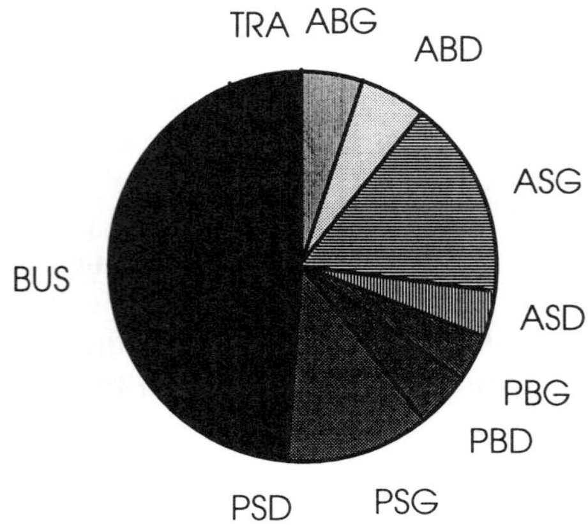

Figure 3.3. Modal shares is the reference year (all periods, pass-km.). BUS, buses; TRA, trains; A, driver alone; P, packed; B, big; S, small; G, petrol; D, diesel.

affecting these vehicles in the base year.[15] However, whereas demand for large cars fueled by petrol increases by 3.3%, total demand for small cars is reduced by 1.7%. Since small petrol-driven cars constitute the largest share of total private vehicles (Figure 3.3), the number of circulating vehicles, including equivalent vehicle units associated with public transport services, declines in equilibrium. This, in turn, causes an increase in average speed, a reduction of generalized transport costs, and a reduction in accidents.[16]

Interestingly, the TRENEN model does not predict higher demand levels for cleaner transport modes, although external costs are fully internalized. This outcome crucially depends on the initial reference situation and, in particular, on the seemingly sub-optimal structure of taxes and subsidies affecting generalized prices in the base year.

3.4.2. *A price realignment experiment*

The TRENEN model identifies a structure of taxes (with endogenous tax revenue) in the presence of external costs. The optimization problem can be usefully decomposed into two steps: (1) an optimal taxation structure is chosen, which maximizes the utility of the representative consumer for a given level of tax revenue, ignoring external costs; (2) external costs are internalized into consumer prices, possibly changing the total amount of taxation. To single out the contribution of each step, it is possible to compare the results obtained from the unconstrained optimization exercise with those from step (1). The latter is called a policy of realignment. It can be formulated, in a simplified way, as:

$$\max_{t_i, \ldots, t_n} V(p_1 + t_1, p_2 + t_2, \ldots, p_n + t_n, Y) \quad \text{s.t.} \quad \sum_i t_i q_i \geq R \tag{3.3}$$

Table 3.1. Changes in macro-variables.

	Realignment	Optimum
Pollution costs	−8.44%	−26.77%
Accident Costs	1.26%	−5.77%
Equivalent income	0.24%	0.30%
Total welfare	0.75%	0.85%

Here we consider a single representative consumer, consuming n goods (where demand levels are q), whose utility is expressed by an indirect utility function V. The latter is assumed to have standard properties of differentiability, and to be negatively related to the prices (inclusive of taxes) and positively related to exogenous disposable income Y. Given a certain target of tax revenue R, the solution of equation (3.3) is a vector of indirect taxes that minimizes the impact on consumer's utility.

First order conditions associated with equation (3.3) are:

$$\frac{\partial V}{\partial t_i} - \lambda \left(q_i + \sum_j t_j \frac{\partial q_j}{\partial t_i} \right) = 0 \tag{3.4}$$

These provide the following optimality condition, obtained from equation (3.4) with the application of Roy's identity:[17]

$$\frac{t_i}{p_i + t_i} \varepsilon_{ii} + \sum_{k \neq i} \frac{t_k}{p_i + t_i} \frac{q_k}{q_i} \varepsilon_{ik} = \frac{t_j}{p_j + t_j} \varepsilon_{jj} + \sum_{k \neq j} \frac{t_k}{p_j + t_{ji}} \frac{q_k}{q_j} \varepsilon_{jk} \tag{3.5}$$

where i and j are two goods or services included in the utility function, and ε indicates own- and cross-price elasticities. From equation (3.5) it easy to verify that, if cross price effects are negligible, two goods in the consumption set are optimally taxed when the tax levels (as fraction of gross prices) are inversely proportional to the demand elasticities. Alternatively, both goods should be subsidized if the exogenously imposed tax revenue R is negative.

The realignment of prices to resource costs and revealed preferences produces some changes in the macro-variables (total pollution and accident costs, equivalent income and welfare as a difference between consumer income and external costs),[18] as shown in Table 3.1. The change in the structure of taxes and subsidies accounts for as much as 88% of the welfare gains obtained with optimal prices in TRENEN. Also, 80% of the variation of equivalent income from the base calibration situation is due to a redistribution of taxes and subsidies.

The equivalent income is higher in the optimum because the level of total indirect taxation is changed. Total tax revenue is indeed negative in the calibration year. Although price distortions within the market are eliminated in the realignment scenario, distortions between consumption of transport services

Table 3.2. Changes in demand quantities.

Transport type	Realignment		Optimum	
	Peak	Off-peak	Peak	Off-peak
Large car				
Single driver				
Petrol	5.19%	7.55%	0.25%	4.23%
Diesel	24.81%	27.71%	19.63%	24.89%
Car pool				
Petrol	4.97%	7.31%	0.29%	4.21%
Diesel	23.76%	26.55%	18.78%	24.03%
Small car				
Single driver				
Petrol	0.88%	3.14%	−4.92%	−0.82%
Diesel	9.40%	12.00%	4.78%	9.44%
Car pool				
Petrol	0.78%	3.02%	−4.68%	−0.65%
Diesel	9.06%	11.51%	4.53%	9.21%
Buses	6.82%	−31.84%	−19.38%	−49.69%
Trains	−3.49%	−44.09%	−27.53%	−58.26%
Other	1.40%	0.26%	−0.80%	0.00%

and consumption of other goods are eliminated only in the optimum. This effect outweights the decrease of consumer welfare due to higher taxes.

Table 3.2 shows how demand for the different transport modes varies in the two regimes. To make it easier to interpret the simulation results, Figure 3.3 shows how total demand for motorized transport modes (passenger-km) is composed in the base year. The presence of external costs in TRENEN causes a general reduction of demand levels. However, demand patterns are already shaped in the realignment scenario, with a drop in the demand for public transport (especially off-peak), and an increase in the demand for large and diesel cars.[19]

3.5. Implicit distributional policies in the transport sector

It is often argued that an unregulated transport market would not ensure the supply of transport services to small communities located in remote areas, or to the poorest population groups. Sometimes the public provision of transport services is seen as the supply of a minimum, guaranteed level of service accessible to all citizens. The level and quality of private transport is also often considered as a signal of available income. In Italy, some taxes are proportional to the car engine power, and other taxes are levied on some luxury cars. This type of discriminatory taxation can hardly be justified by looking at the transportation sector in isolation. For example, although large cars may pollute more, the additional tax burden is much more than proportional to the level of additional pollution. Tax discrimination on the consumption of transport services seems, therefore, to be used as an alternative instrument to progressive income taxation. From this point of view, to the extent that public transport is mainly used by relatively poorer people, subsidies assigned to public transport and taxes imposed on private transport may be the outcome of some kind of distributional policy.

In this section, we shall investigate whether implicit distributional policies may explain, to some extent, the large gap that is observed between market prices of local transport services and shadow prices arising from resource costs and consumer preferences. One way to do this is by assuming that, instead of the utility of a representative consumer, a Bergson–Samuelson social welfare function is maximized by a central planner. The social welfare is in this case a function of the utility levels associated with different population groups:

$$W(v_1, v_2, \ldots, v_n) \tag{3.6}$$

If, for example, two groups are considered and the social welfare function takes a linear form, the optimization problem (equation 3.3) can be restated as:

$$\begin{aligned} &\max_{t_i, \ldots, t_n} \beta^a v^a(p_1 + t_1, p_2 + t_2, \ldots, p_n + t_n, Y^a) \\ &+ \beta^b v^b(p_1 + t_1, p_2 + t_2, \ldots, p_n + t_n, Y^b) \\ &\text{s.t.} \quad \sum_i t_i(q_i^a + q_i^b) \geq \bar{R} \end{aligned} \tag{3.7}$$

Optimality conditions for the solution of equation (3.7) can be written in a compact and convenient way, by applying again Roy's identity and by recalling that: (1) when indirect utility functions v are money metric, that is expressed as equivalent income, the marginal utility of income is equal to one; (2) the objective function can be freely rescaled and this allows us to normalize to one the Lagrangian multiplier associated with the constraint. Using Roy's identity, first order conditions can then be expressed as:

$$\beta^a q_i^a + \beta^b q_i^b = R_i \tag{3.8}$$

where:

$$R_i = q_i^a + q_i^b + \sum_j t_j \left(\frac{\partial q_j^a}{\partial t_i} + \frac{\partial q_j^b}{\partial t_i} \right) \tag{3.9}$$

The left-hand side of equation (3.8) expresses the social welfare cost, in money terms, of raising indirect taxes on the good i, and the right hand side is the marginal increase in tax revenue. The system of optimality conditions (equation 3.8) can be used to infer values for the unknown distributive weights if, generalizing the calibration approach, it is assumed that tax levels observed in a certain base year are determined by an implicit social welfare maximization. This problem is known in the literature as the inverse optimum problem (Ahmad and Stern, 1984).

To illustrate this point, consider an example with two consumers and two goods. Suppose that the second good is not taxed and is consumed only by the second consumer. The optimality condition associated with this good would then be very simple:

$$\beta^b q_2^b = q_2^b \tag{3.10}$$

Clearly, the only value of the distributive weight for the consumer b which satisfies equation (3.10) is one. Suppose now that the first good is consumed by both consumers, that there are no cross price effects, and that the price derivative of the aggregate demand is equal to -1. The corresponding optimality condition is in this case:

$$\beta^a q_1^a + \beta^b q_1^b = q_1^a + q_1^b - t_1 \tag{3.11}$$

Using equation (3.10) and rearranging terms, we get:

$$\beta^a = 1 - \frac{t_1}{q_1^a} \tag{3.12}$$

A higher tax on the first good (and, correspondingly, a lower level of demand) therefore indicates a smaller relative distributive weight of the consumer a. Indeed, although consumer b is also negatively affected by a higher tax, the impact on this utility would be lower, because consumption is not completely concentrated on the first good.

This example can be easily generalized to the case of n goods and n consumers. Optimality conditions (equation 3.8) give raise to a linear system, whose solution is a set of 'revealed' distributive weights. In matrix terms, this is:

$$\mathbf{Qb} = \mathbf{r} \Rightarrow \mathbf{b} = \mathbf{Q}^{-1}\mathbf{r} \tag{3.13}$$

where $\mathbf{b}$ is a vector of distributive weights, $\mathbf{Q}$ is a $(n \times n)$ matrix of demand levels by consumer groups at a certain calibration year, and $\mathbf{r}$ is a vector of estimated marginal tax revenue.

When consumption items are less than the number of consumers, the vector of distributive weights is not unique. If, on the other hand, the number of consumers is less than the number of consumed goods and services, the matrix $\mathbf{Q}$ is rectangular and therefore cannot be inverted. Alternatively, one can use a 'generalized inverse' matrix $\mathbf{Q}^* = (\mathbf{Q}'\mathbf{Q})^{-1}\mathbf{Q}$, computing $\mathbf{b}$ via:

$$\mathbf{b} = (\mathbf{Q}'\mathbf{Q})^{-1}\mathbf{Qr} \tag{3.14}$$

which amounts to a least squares estimation of $\mathbf{b}$, where optimality conditions are not (in general) simultaneously satisfied, but a bias error is minimized (Christiansen and Jansen, 1978).

This is precisely the case which is considered here, because we disaggregate the structure of the demand for urban transport services in Bologna by two income groups.[20] First, implicit distributive weights are derived from the current tax structure using equation (3.14). Second, the TRENEN optimization procedure is repeated, by considering two representative consumers who enter the welfare objective function in a non-symmetrical way.

Data requirements for the implementation of this procedure are quite severe. In principle, one would need data on consumption of transport services in a certain city disaggregated by income groups. These data are not normally available, but indirect information can be used to infer plausible consumption values. In our case, we combined three information sources: (1) aggregate demand levels, used in the single consumer version of TRENEN, (2) data on households expenditure at the national level, specified by income groups and referred to some classes of transport services, (3) the relative size of income groups, where we have divided all households into two groups, depending on whether monthly consumption expenditure does or does not exceed 3 millions lire. This information, supplemented by educated guesses on the demand structure, provided us with the necessary data set for the estimation of the implicit distributive weights.

As expected, given the relatively larger consumption share of public transportation and small cars by the poor consumer group, the estimated weight for this group was higher than that of the rich: the distributive weight for a representative poor consumer is 1.33 and the weight for a rich consumer is 0.6. The gap between the two weight parameters tends to enlarge when demand for transport services is, on average, more elastic and/or when the relative size of the rich group is smaller.

In order to include distributive weights in the TRENEN model the welfare objective (equation 3.2) has been modified in the following way:

$$W_s = \left(\beta^p q P_c^p \frac{Y_c^p + \Delta T_s^p}{P_s^p} + \beta^r (1-q) P_c^r \frac{Y_c^r + \Delta T_s^r}{P_s^r} \right) - E_s \qquad (3.15)$$

where p and r refer to the poor and rich population groups, and q is the proportion of poor consumers in total population. For equation (3.15) to produce comparable results with the single representative consumer case, it is necessary that

$$q Y_0^p + (1-q) Y_0^r = Y_0 \qquad (3.16)$$

In other words, we require that the initial income of the representative consumer in equation (3.2) is a weighted average of the two incomes, for the poor and the rich consumer.

Unconstrained maximization of equation (3.15) gives rise to a new set of taxes, prices and demand quantities. Taxes change, in comparison to the base case, both in terms of levels and in terms of composition. It is possible to see how the overall tax burden changes by comparing the two objective functions. To this end, consider what would happen if the equivalent income for the single

Table 3.3. Changes in macro-variables (optimization with distributive weights)

	Large income inequality		Small income inequality	
	Rich	Poor	Rich	Poor
Pollution costs	−62.34%	−13.57%	−62.29%	−13.64%
Accident costs	−4.09%	−6.18%	−3.98%	−6.26%
Equialent income	0.21%	0.32%	0.24%	0.29%

consumer would be multiplied by a factor greater than one. Since objective functions can be freely rescaled (the function can be divided by the scaling factor), this would amount to a proportional reduction of externality costs and, consequently, to optimal tax levels. Conversely, a multiplying factor less than one would give rise to higher tax levels.

Following this kind of reasoning, one could compare the equivalent income in equation (3.2) with the bracketed term in equation (3.15). Despite equation (3.11), the two elements are not, in general, equal, because of the presence of distributive weights. The composite surplus indicated in equation (3.15) would be lower than the single consumer surplus (so, taxes would, on average, be higher) if the income of the poor consumer is less than 80.8% of the income of the rich.

In order to assess the sensitivity of the model results to income distribution assumptions, we ran two different simulation exercises, both using equation (3.15) in the first, income of the poor was set at 70% of the income of the rich, in the second, the income of the poor was set at 90%. The results are summarized in Table 3.3. Values of macro-variables for the single representative consumer (Table 3.3) are always in the interval of values estimated for the two consumers case, with the exception of the equivalent income when income levels are not very different. Because of the higher weight assigned to the poor consumers group, the gain in equivalent income is always larger for the representative poor, but there is less discrimination when income inequality is initially limited. A very significant share of the total reduction in pollution is due to the reduced consumption of the rich. This is because taxes on transport modes preferred by the poor cannot be raised too much, given the existence of distributional policy objectives.

The structure of generalized prices, taxes and demand levels is not so different from the structure stemming from the unconstrained optimization, single consumer case. This can be seen either on the side of optimal tax levels, or on the side of consumption patterns. Tables 3.4 and 3.5 show the composition of demand for the poor and rich consumers group, under the two alternative assumptions of income distribution. Comparing Tables 3.2, 3.4 and 3.5, the following conclusions can be drawn. First, the demand pattern for both the rich and the poor is quite similar to that of the representative consumer. For

Table 3.4. Changes in quantities (optimization with distributive weights, large income inequality)

Transport	Rich		Poor	
	Peak	Off-peak	Peak	Off-peak
Large car				
Single driver				
Petrol	−1.82%	2.10%	0.23%	4.39%
Diesel	16.78%	21.79%	19.21%	24.52%
Car pool				
Petrol	−2.43%	0.94%	0.32%	3.75%
Diesel	15.51%	19.76%	18.78%	23.10%
Small car				
Single driver				
Petrol	−5.43%	−1.81%	−3.43%	0.41%
Diesel	3.76%	8.00%	5.97%	10.43%
Car pool				
Petrol	−5.19%	−2.43%	−2.38%	0.37%
Diesel	2.93%	6.49%	5.98%	9.55%
Buses	−19.51%	−49.32%	−19.20%	−48.60%
Trains	−27.66%	−59.12%	−27.38%	−58.54%
Other	−0.73%	0.06%	−0.80%	0.09%

Table 3.5. Changes in quantities (optimization with distributive weights, small income inequality)

Transport	Rich		Poor	
	Peak	Off-peak	Peak	Off-peak
Large car				
Single driver				
Petrol	−1,71%	2,21%	0,15%	4,30%
Diesel	16,91%	21,92%	19,12%	24,42%
Car pool				
Petrol	−2,32%	1,05%	0,24%	3,67%
Diesel	15,63%	19,89%	18,68%	23,00%
Small car				
Single driver				
Petrol	−5,32%	−1,70%	−3,51%	0,32%
Diesel	3,88%	8,11%	5,87%	10,34%
Car pool				
Petrol	−5,09%	−2,33%	−2,47%	0,29%
Diesel	3,05%	6,60%	5,89%	9,46%
Buses	−19,42%	−49,27%	−19,26%	−48,64%
Trains	−27,58%	−59,08%	−27,44%	−58,57%
Other	−0,63%	0,16%	−0,90%	−0,01%

example, the reduction in demand for large cars by the poor (large inequality) is almost the same as shown in Table 3.2, whereas demand for small cars and public transport is slightly higher. Second, reductions in consumption levels are always larger and increments are always smaller for the rich, with the exception of walking, cycling and other transport modes. Again, the discrimination between rich and poor is less when there is less income inequality. Finally, we can still observe a very large reduction in the demand for public transport modes. This is because subsidies are eliminated even in the two consumer case: the introduction of distributive weights is not sufficient to justify the existence of subsidies, as demand patterns for the two income groups are not very different.

3.6. Concluding remarks

Economic theory states that optimal prices should include all internal and external costs. When a market is close to perfect competition external costs can be added to market prices by Pigouvian taxes, so that consumption levels are reduced in proportion to generated externalities. The market for urban transport services is, however, far from being perfectly competitive. The imposition of an optimal price structure in this market does not necessarily bring about a reduced consumption for transport modes associated with higher external costs. Distortions in the urban transport market may be due to political constraints and inefficient regulation, but also to policy objectives which are external to the market. In general, it is difficult to ascertain the relative magnitude of these two effects.

In this paper, we used an optimization model to determine optimal price levels for urban transport services in the city of Bologna. The price distortions are so severe that optimal prices are higher for public transport and lower for automobiles, even when external costs are internalized. Furthermore, most of the welfare gain does not come from internalizing congestion and pollution costs, but rather from realigning prices closer to resource costs. Despite lower prices for automobiles (on average), pollution costs in the optimal scenario decline because of a better allocation of transport consumption between periods of the day and transport modes, as well as because of the introduction of some pollution abatment technologies (like catalytic converters).

We ran the model with one and two representative consumers. In the second case, we focused on a specific external policy objective (real income distribution), deriving implicit distributive weights from current fiscal policies. Optimal prices were then computed while taking into account some exogenous distributional targets. Our simulation results, however, suggest that distributional policies do not explain the presence of price distortions in the market for urban transport services.

Inefficient regulation, the existence of other policy objectives and constraints, or of other externalities (even positive ones) could provide alternative explanations: these should be taken into account to formulate an effective and politically feasible environmental policy in the field of urban transportation. If some factors affecting the current structure of prices, subsidies and taxes, are forgotten and green taxes are introduced only on the basis of marginal environmental costs, distorsions in the market may be exacerbated, causing welfare losses and increased inefficiency.

Notes

1. This paper is partly based on some preliminary results obtained by a model developed within the research projects Trenen (Joule II) and Trenen II Stran (Transport) of the European Commission (DG XII). Research assistance from Stefania Pasquon, Laura Valentini, and Silvio Pancheri is gratefully acknowledged. Detailed comments by Kenneth Small and an anonymous referee helped to improve earlier versions of this work. However, the usual disclaimer applies.
2. Attempts have recently been made to assign new powers to regional governements in the field of transportation (e.g. regional subsidies and regionalized railway fares).
3. For more details on the TRENEN model (inter-regional and urban), see the final research report (European Commission, DgXII, 1995), which also illustrates various simulation experiments.
4. Implemented versions of TRENEN may also differ about the treatment of tax revenues. In some cases, tax revenue is accounted for separately and weighted in the social welfare function using a parameter expressing 'the shadow cost of public funds' (see Chapter 2). In other cases, tax revenue is redistributed to the representative consumer. For a discussion about the implications of the two assumptions on the model results, see Roson, 1995.
5. More recent versions of the urban model include an exogenous demand for freight transportation. Freight is not included in this chapter.
6. Therefore, these indexes are not affected by approximation biases, as in Paasche or Laspeyres indexes. Because of the costant return to scale assumption adopted in TRENEN, unit price indexes are also unaffected by changes in demand levels.
7. Some fixed costs are made dependent on total passenger-km. This may be justified in this context, because the model is an aggregate one, and a representative consumer is used to model consumption choices of thousands of individuals. A new quasi-dynamic version of the model, however, provides a better distinction between fixed and variable costs.
8. Parameters are estimated by interpolating a number of average speed/traffic volume observation points. These are provided by a geographically detailed network equilibrium model, using different origin-destination flow matrices.
9. The model can easily simulates cases in which no tax discrimination is possible, by simply adding specific constraints to the optimization process.
10. In general, this does not allow one to estimate all parameters in the model. The number of estimable parameters equals the number of optimality conditions that must be satisfied exactly in equilibrium. The remaining parameters (in this case, the elasticities of substitution) are set exogenously.
11. Quite often, the possibility of making a choice between two transport modes is limited: for example, in the choice between public and private transport, one could have to consider that a public service may not reach a certain location, or may not be available in certain hours of the day.
12. There are two factors determining lower environmental costs. First, demand for small petrol-driven cars slightly declines, in addition to the drop in the demand for public transport. Use of

private cars does not increase very much, because of the large share of small cars. Second, in equilibrium, large petrol-driven cars are equipped with catalytic converter (cc). In the calibration it is assumed that converters are not installed in any car. The substitution elasticity between the alternatives 'with cc' and 'without cc' is very high, so in equilibrium either there is a catalytic converter on all cars of a certain type, or not. The model finds it optimal to install converters only on large cars using petrol. These vehicles are taxed according to their social marginal cost, which is significantly smaller than the calibration one.

13. Note that generalized prices include marginal resource costs (unchanged), time costs (reduced because of a lower level of circulating vehicle units) and taxes (increased, on average, because of the internalization). The model simulates the creation of an additional (redistributed) tax revenue.
14. A second-best situation in which subsidies to public transport may emerge is when congestion or other external costs are relevant and there exist an upper bound for taxes on private transport. However, no such bound is imposed here.
15. See also the remarks in the Appendix.
16. Equivalent vehicle units decline by 3% in the peak and by 1.8% in the off peak period. Average speed increases only by 0.5% and 0.4%, respectively. The value of time adopted in all the simulation exercises is 4.77 ECU/h/passengers of cars and 3.61 ECU/h/passengers of public transport and non-motorized modes.
17. Roy's identity states that:

$$q_i = -\frac{\partial V/\partial p_i}{\partial V/\partial Y} = -\frac{\partial V/\partial t_i}{\partial V/\partial Y}$$

This holds for every good *i*. Combining first order conditions for two or more goods eliminates the marginal utility of income and the Lagrangean multiplier, giving rise to equation (3.5).

18. Externalities are not taken into account to find optimal taxes in the 'realignment' scenario. However, external costs can be computed *ex post* on the basis of simulated demand levels. Although pollution costs decrease in both cases, the composition of pollutants is different. In the optimum, VOC are 22.5% less than in the realingment, NO_x pollutants are reduced by 7.4%, but CO_2 emissions are reduced only by 1.1%. Also, congestion externalities are not considered in the 'realignment' scenario, but can be computed *ex post*.
19. In 1991, a special additional tax ('superbollo') was in effect on diesel vehicles. This special tax was subsequently abolished.
20. One may argue that it is not correct to infer implicit distributive weights by looking only at indirect taxation in the transport sector, and that the whole tax system should instead be taken into account. This would be irrelevant if a consistent fiscal policy applies the same distributional criteria in all sectors. Although this is unlikely to occur in the real world, in this exercise we only want to single out distributional objectives in the transport taxation, possibly including inconsistent ones.

References

Cobello, F., 1994, Il trasporto pubblico locale nel Veneto (Local public transport in the Veneto region), Economia Pubblica, 3, 123–141.

Conto Nazionale dei Trasporti, published yearly by the Italian Ministry of Transportation.

Amhad, E. and N. Stern, 1984, The theory of reform and Indian indirect taxes, Journal of Public Economics, 25, 259–298.

Christiansen, V. and E.S. Jansen, 1978, Impicit social preferences in the Norwegian system of indirect taxation, Journal of Public Economics, 10, 217–245.

European Commission DG XII, 1995, TRENEN – Final Report, Brussels: Research Project Joule II.

Keller, W.J., 1976, A nested CES-type utility function and its demand and price index functions, European Economic Review, 7, 175–186.

Mayeres, I., S. Ochelen and S. Proost, 1996, The Marginal External Costs of Transport, Public Economics Research Paper N. 51, Leuven: CES, Katholieke University.

Ministero dei Trasporti, 1994, Radiografia delle FS 1992, Roma: DGPOC.

Roson, R., 1995, Fiscal Policies and Externalities for Complex Technologies: the Transport Case, GRETA working paper n.95.04.

3A. Appendix: data on external environmental costs

The simulation exercises described in this chapter were carried out with a preliminary version of the TRENEN model, which uses only aggregate information regarding pollutants and vehicle types. Only three classes of pollutants are considered, and emissions are assumed to be invariant with vehicle size, period of the day (congestion), and carried passengers. The model, however, distinguishes between vehicles equipped with catalytic converter (CC), and those which are not (NC). Table 3A.1 shows emissions by vehicle type, in g/vehicle-km, the marginal social cost of each pollutant, in ECU/g, and the total external environmental costs.

A more recent version of the model uses better data on environmental costs. Using the survey by Mayeres et al. (1996), emissions are distinguished by vehicle size, number of passengers, and traffic congestion (that is, period of the day). Table 3A.2 shows the new parameter values.

Three other classes of pollutants are taken into account. The inclusion of particulates (PM) makes total environmental costs quite different for diesel cars and buses (which are mainly diesel in Italy). This, of course, brings about much higher optimal prices for diesel vehicles, and lower demand levels, than those suggested by earlier simulation exercises. Nonetheless, since diesel cars constitute a small proportion of total circulating vehicles and most of the price change for buses is still due to the elimination of subsidies, results obtained with the new data set are not significantly different (at the aggregate level) from those illustrated in this work.

Table 3A.1. Marginal pollution and costs by vehicle type (data set I).

	NO_x	CO_2	VOC	Exernal costs
Valuation	0.002	2E-05	0.0045	
GAS.NC	0.56	181.44	3.5	0.0201
GAS.CC	0.1	192.78	0.4	0.0055
DIE.NC	1	163.8	0.5	0.0072
DIE.CC	0.1	131.04	0.5	0.0048
BUS	21.3	1008	5.3	0.0846
TRA[a]	0.8571	336.6	0	0.0078

[a]Data refer to a hypothetical vehicle carrying the same number of passengers as a bus.

Table 3A.2. Marginal pollution and costs by vehicle type (data set II).

	NO_x	CO_2[a]	VOC	CO	PM	SO_x	
Valuation	0.0138	0.0077	0.0029	1E−05	0.0832	0.0952	External costs
PEAK.ALON.GAS.SML.NC	1.559	0.167	0.365	2.916	0	0	0.0239
PEAK.ALON.GAS.BIG.NC	2.195	0.225	0.514	4.246	0	0	0.0336
OFFP.ALON.GAS.SML.NC	1.388	0.142	0.297	1.803	0	0	0.0211
OFFP.ALON.GAS.BIG.NC	1.924	0.175	0.418	2.632	0	0	0.0291
PEAK.POOL.GAS.SML.NC	1.723	0.18	0.39	3.063	0	0	0.0263
PEAK.POOL.GAS.BIG.NC	2.394	0.236	0.545	4.423	0	0	0.0365
OFFP.POOL.GAS.SML.NC	1.506	0.15	0.315	1.894	0	0	0.0229
OFFP.POOL.GAS.BIG.NC	2.052	0.183	0.441	2.782	0	0	0.031
PEAK.ALON.GAS.SML.CC	0.283	0.167	0.06	0.33	0	0	0.0054
PEAK.ALON.GAS.BIG.CC	0.399	0.225	0.085	0.48	0	0	0.0075
OFFP.ALON.GAS.SML.CC	0.252	0.142	0.049	0.204	0	0	0.0047
OFFP.ALON.GAS.BIG.CC	0.349	0.175	0.069	0.298	0	0	0.0064
PEAK.POOL.GAS.SML.CC	0.313	0.18	0.064	0.346	0	0	0.0059
PEAK.POOL.GAS.BIG.CC	0.435	0.236	0.09	0.5	0	0	0.0081
OFFP.POOL.GAS.SML.CC	0.274	0.15	0.052	0.214	0	0	0.0051
OFFP.POOL.GAS.BIG.CC	0.373	0.183	0.073	0.314	0	0	0.0068
PEAK.ALON.DIE.SML.NC	0.201	0.136	0.015	0.33	0.066	0.13	0.0217
PEAK.ALON.DIE.BIG.NC	0.322	0.168	0.032	0.67	0.108	0.16	0.0301
OFFP.ALON.DIE.SML.NC	0.157	0.099	0.009	0.26	0.041	0.09	0.0149
OFFP.ALON.DIE.BIG.NC	0.228	0.126	0.016	0.56	0.063	0.12	0.0208
PEAK.POOL.DIE.SML.NC	0.206	0.144	0.014	0.34	0.068	0.13	0.022
PEAK.POOL.DIE.BIG.NC	0.331	0.178	0.028	0.67	0.111	0.17	0.0314
OFFP.POOL.DIE.SML.NC	0.161	0.105	0.009	0.28	0.047	0.1	0.0165
OFFP.POOL.DIE.BIG.NC	0.235	0.134	0.016	0.55	0.066	0.12	0.0212
PEAK.ALON.DIE.SML.CC	0.201	0.136	0.015	0	0.0066	0.13	0.0168
PEAK.ALON.DIE.BIG.CC	0.322	0.168	0.032	0	0.018	0.16	0.0226
OFFP.ALON.DIE.SML.CC	0.157	0.099	0.009	0	0.0041	0.09	0.0119
OFFP.ALON.DIE.BIG.CC	0.228	0.126	0.016	0	0.0063	0.12	0.0161
PEAK.POOL.DIE.SML.CC	0.206	0.144	0.014	0	0.0067	0.13	0.0169
PEAK.POOL.DIE.BIG.CC	0.331	0.178	0.028	0	0.011	0.17	0.0231
OFFP.POOL.DIE.SML.CC	0.161	0.105	0.009	0	0.0047	0.1	0.013
OFFP.POOL.DIE.BIG.CC	0.235	0.134	0.016	0	0.0066	0.12	0.0163
PEAK.BUS	19.81	0.9483	1.7076	0.4338	3.916	0.9037	0.6975
OFFP.BUS	14.428	0.6682	1.1748	0.2205	1.811	0.6369	0.419
PEAK.TRA	0.8571	0.3366	0.0083	0.0862	0	1.3327	0.1413
OFFP.TRA	0.8571	0.3366	0.0083	0.0862	0	1.3327	0.1413

PEAK, peak period; OFFP, off-peak; GAS, petrol-driven; DIE, diesel; SML, small can; BIG, large car; ALON, single person in car; POOL, pooled transport.

CHAPTER 4

Environmental effects and scale economies in transport modelling: some results for the UK[1]

John Peirson and Roger Vickerman

4.1. Introduction

Much interest is currently being focused on the extent to which the private sector can assume many of the functions in transport which have traditionally been the preserve of the public sector. Thus, we are experiencing concerted attempts to privatize the provision not only of bus services, but also of entire railway networks. The attempt at involving the private sector is going beyond the provision of services to the provision of the infrastructure itself. Initially, this has involved using the private sector to generate investment funds for discrete new projects. The public sector, facing serious debt constraints, can no longer provide such funds as cheaply as previously and moreover the private sector has been argued to be potentially more efficient in managing the development of such projects. However, current plans go beyond this minimal involvements, for example, the wholesale privatization of complete networks, such as Railtrack, the owner of rail track infrastructure in the UK, which was sold to the private sector in May 1996.

Along with this interest in introducing the market to achieve the advantages of improved efficiency, competition and improved service quality, there is increased awareness of the market failures associated with transport. The traditional argument for the involvement of the public sector in transport was based on the existence of scale economies, which led to a natural monopoly, particularly in the provision of infrastructure. This led in turn to a need for regulation. At the other extreme, where large scale economies were not present and entry costs were low, the opposite tendency would be observed, usually referred to as destructive competition. Although prices would be reduced in the short term, this would usually be to levels below long run marginal cost which would preclude new investment. Despite the short-term gains, the long-term effect would be against the interests of consumers and would require regulation.

R. Roson and K.A. Small (eds.), Environment and Transport in Economic Modelling, 61–75.

Moreover, there were seen to be quality issues in transport provision which would require regulation by the public sector. In particular, safety requirements would necessitate some form of quality threshold which may be a more satisfactory form of control on market entry than quantity controls. Increasingly, however, these quality issues have become more concerned with the general environmental impact of transport. The nature of this environmental concern has also changed. It started with a general concern over congestion and pollution, but this has led more recently to an increasing focus on the detailed evidence concerning different types of external effect. These include global warming effects, the impact of local air pollution on certain chronic medical conditions, especially through particulates, the effects of noise, and the costs of congestion and accidents.

In this paper, we bring together the major themes of scale economies and environmental effects in a broad aggregate model of transport demand in the UK. The model is developed at a fairly simple level, but demonstrates some useful insights into issues which are too frequently overlooked. This enables us to explore the relationship between scale economies, as determinants of differences in the internal costs and hence the supply price of different modes, and their environmental impacts, as determinants of external costs. The costs of these various external effects are brought together on a common basis for each mode (see Peirson et al., 1995). The model is calibrated in order to produce optimal inclusive prices in both a short-run situation, in which capacity is given, and the long run, in which investment in additional capacity can take place (see Peirson and Vickerman, 1996, 1997, for fuller details of this model).

4.2. Scale economies and external effects in transport

The basic economics of transport identifies two critical analytical points: the existence of scale economies which, combined with lumpy investment, lead to critical questions of capacity, and the importance of congestion as an imperfect rationing device. Scale economies can exist both in the provision of infrastructure, and in the purchase and operation of vehicles. The view that there are also substantial economies of scale in the size of fleets has been challenged, along with some of the assumptions of scale economies in the operation of larger vehicles (especially those connected with labour costs). However, there is a growing concern about the extent of network economies, which could be lost in the privatization and consequent break up of, for example, national rail networks.

The exact extent of scale economies in the provision of transport is still the subject of considerable debate. This is of critical importance in understanding the optimal provision of transport when taking external effects into account. The core issue is the extent to which scale economies in transport are complements to, or substitutes for, lower external effects. In other words, scale economies could, on the one hand, be used as a justification for maintaining modes of transport with large external effects because the scale economy effect reduces the full social costs to the community. On the other hand, if environmentally less damaging modes have higher internal costs making them expensive to provide, but the potential for securing substantial scale economies, can this be used to justify public subsidy in the interest of exploiting the scale economies (the natural monopoly argument)?

A detailed analysis of this argument is given in Peirson and Vickerman (1996). The essence of this relationship is a recognition that the imposition of full social costs will lead to a rise in the price of all modes, but by different relative amounts. Thus, given the existence of non-zero cross-elasticities of demand, each mode will experience, as well as a price effect, an (albeit small) income effect, which arises from the movement in the price of a substitute mode. Since the optimal price for allocating a given capacity will be the short-term marginal social cost, in the presence of large-scale economies this will produce an operating deficit since marginal costs are below average costs. The question is then whether the 'tax' revenues raised to pay for the external costs will be sufficient to cover these deficits. If the shift of demand from one mode (with large external effects) to another (with smaller external effects) leads to capacity constraints in the latter, there is, in addition, a case for using the tax revenue from the former to fund the investment in the latter in order to achieve a long run social optimum. This suggests a theoretical case for a self-financing transport system, though whether the tax revenues are sufficient to finance the deficits and the investment is an empirical question which will depend on the size of the scale economies and the elasticities, both own price and cross-price. The use of a multimodal model here takes us beyond the simple example of the theoretical case for a self-financing road authority which is exactly funded by optimal congestion charges (Newbery, 1994).

These arguments concerning public subsidy to exploit scale economies raises a further issue. One characteristic of transport is that it tends (in the case of most modes) to involve lumpy investments. Changes of capacity imply relatively large discrete changes. In the short term, transport systems are characterized by either surplus capacity or inadequate capacity and excessive congestion. Thus, any attempt to shift traffic from one mode to another can result in various different situations depending on the degree of capacity utilization in each mode. If two modes are operating at less than full capacity, any shifting of demand from one to the other on environmental grounds has little significance and the welfare gain derives simply from the environmental improvement. If the losing mode is operating at no greater than full capacity, but the potentially gaining mode is at full capacity, then any shift towards the capacity constrained mode on environmental grounds is likely to be costly in terms of necessary new capacity. If, however, the losing mode is operating at above full capacity, then there may be a gain from any reduction in its over-utilization apart from any environmental gain. However, any such shift which leads to a need for new investment in one mode whilst reducing the demand for another mode and leading to its operation at less than capacity may create greater burdens in the losing mode which now has to maintain the redundant capacity, since it is typically not possible to reduce capacity costlessly.

This point becomes particularly clear in the treatment of urban and inter-urban travel. In the former, the whole network approach has dominated both within and between modes. In the latter, rather less interdependence has been assumed, leading to difficulties in the measurement of congestion or its impacts. Furthermore, approaches have tended to be oriented more towards the analysis of roads as a single mode which might produce misleading results if the cross-effects of imposing charges are not considered (Newbery, 1988, 1990, 1994). However, it is in inter-urban travel where congestion is rapidly becoming a serious problem and where policy attention is being directed to find possible solutions. This is particularly the case where the question arises as to the likely traffic generating characteristics of new road investments. The argument here is essentially one of identifying genuinely newly generated traffic from that redistributed over the network (Department of Transport, 1994a,b). Much inter-urban travel takes place in a competitive market between road, rail and air (depending on distance and origin and destination). Correct relative pricing is important as a pre-requisite to correct investment policy (regardless of who is investing, public sector or private agencies).

4.3. The model

In order to investigate these issues further, we consider an aggregate model, as the presently available UK data precludes the construction of network models which allow for differences in the variables across the different parts of the network. We use a simple demand function calibrated on existing price and income elasticities. The supply side includes the internal operating costs of each mode and, in the long run the costs of the provision of infrastructure. To this are added the various external costs associated with that mode. The model then equilibrates demand and supply in the short run at a price equal to short run marginal social costs (Figure 4.1).

In the short term, the price facing the consumer is taken as the long-run marginal private costs plus marginal external costs. Infrastructure costs are excluded since in the short run they are sunk costs and in the long term there is a problem of double counting, since in equilibrium the marginal cost of congestion must equal the long-term marginal cost of providing additional capacity to reduce congestion. All the external costs are monetized at their marginal value and included in the inclusive price as a Pigovian tax. The model leads to a rebalancing of mode use, and changes in external and internal costs. The solution for equilibrium prices and demand occurs within the model for both short and long term.

In the long term, allowance is made for optimal investment in new capacity.

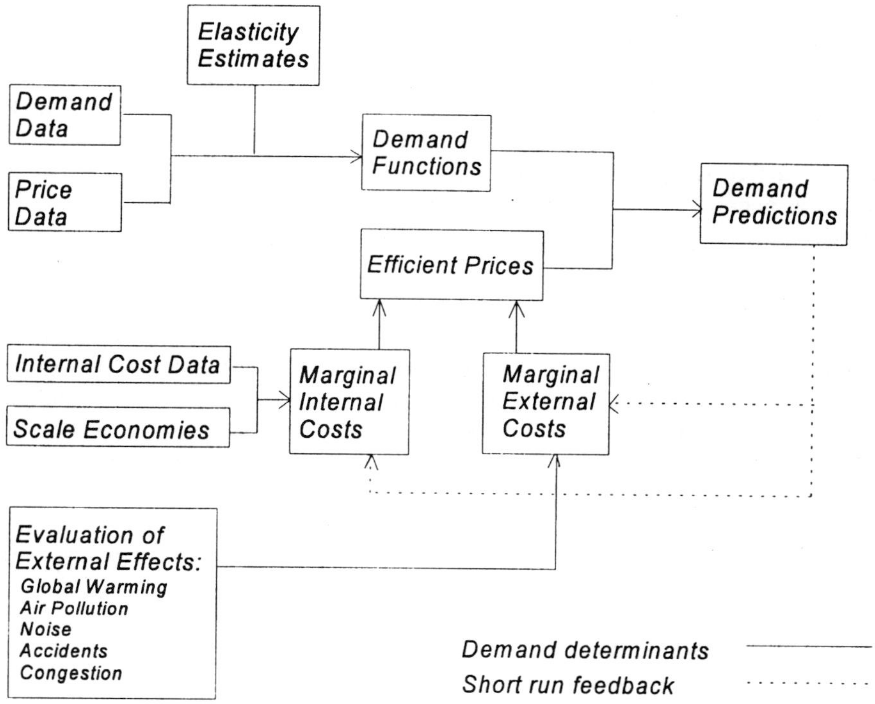

Figure 4.1. Structure of short-term model.

A further iteration is added to the model shown in Figure 4.1 to allow for any new investment required to obtain an optimal level of demand in the long term. This optimal level is also solved within the model. Thus the level of demand is used to assess the optimal level of capacity and hence the investment required. Prices then adjust to reflect the implied level of marginal internal and external costs at this new level of capacity. The initial simple approach assumes constant returns to scale so that long-term marginal private costs are equal to long run average costs for that mode. In a second stage, we allow for any scale economies achievable by the expansion of use of a particular mode. This sets long-term marginal private costs at a fixed value set appropriately lower than average costs, following detailed discussion with the various operators. This constancy of long-term marginal costs ensures the existence of a unique equilibrium.

The model has been developed for two different cases to represent the differing situations in urban (London) and inter-urban travel. Given the aggregate nature of the model, it is not unreasonable to assume that, at least in an urban situation, traffic has a generally good knowledge of the network, which is relatively dense and thus the level of use of the networks is fairly constant over its major links. This is less realistic for the inter-urban case. Here we have considered just the inter-urban sections of journeys; data limitations make it difficult to assign correct values for a complex journey involving both urban and inter-urban sections. We have maintained the assumption of network neutrality, essentially that there is no significant variation in the level of external effects (particularly congestion) at any point on the network. This is known to be less applicable to, for example, the motorway network in the UK, but the modelling of congestion on such roads is difficult and would detract from other properties of the model to be tested here.

In both cases, only passenger travel is included: we assume that freight traffic is essentially fixed at a price which is determined by factors outside the transport network, but more importantly that there are no significant cross-effects between the passenger and freight markets. This is a rather over-simplified assumption, clearly they compete for the use of infrastructure, but it has enabled us to gain experience with the model before introducing this complication.

The demand model used takes the form:

$$D_i = \alpha_i Y^{\gamma_i} \prod_{j=1}^{n} [P_i]^{\beta_{ij}} \tag{1}$$

where D_i = demand for the i_{th} transport mode in passenger-km, P_i = inclusive price of i_{th} transport mode per passenger-km, Y = income, β_{ij} = price elasticity of i_{th} demand with respect to j_{th} price, γ_i = income elasticity of i_{th} demand.

In equation (4.1) the dependent variable is in terms of passenger-km, the price is calculated as described below and the income effects were estimated by assuming that in the long term income growth would be 2.5% per annum, while in the short term income was assumed to be constant. In the short term, price elasticities β_{ij} are smaller than in the long term.

A review of the literature (see Peirson et al., 1994a, for full details) yielded estimates of short- and long-term own and cross price elasticities and income elasticities. Information on transport demand was obtained from Department of Transport (1991, 1992, 1993a, 1993b) and current price data on internal prices was taken from the Department of Transport (1981 plus revisions; 1993a) and British Railways Board (1993) and through detailed discussions with operators. All data relate to the year 1991. Given the values of the demand and independent variables, plus elasticities β_{ij} and γ_i, equation (4.1) was calibrated – α_i can be derived from the equation and then used to estimate the impacts on D_i of changing values of P_j.

The values of P_j depend on both external estimates of cost and the model itself which adjusts price and demand into equilibrium. Thus, as traffic volumes change, the levels of congestion, accidents etc change and hence the value of P_j. The costs required to estimate these inclusive prices depend on the internal costs of the various modes referred to above, and their external impacts. Though some of these external costs have been estimated for the UK in the past (Newbery, 1987, 1990; Pearce, 1993), this study has brought together previous estimates on a consistent and appropriate basis. Full details of the estimation of these are given in Peirson et al. (1994a, 1995). This data has been revised in line with more recent estimates (e.g. Maddison et al., 1996), a full discussion of these and discussion of the principles underlying the approach used is given in Peirson and Vickerman (1998).

Information on economies of scale was taken from surveying the limited information available for the British Isles and detailed discussions with operators. There is a debate on the extent of economies of scale in transport (see, for example, Button, 1993). To allow for this uncertainty, we experimented with three possible scenarios: one of constant costs and two of decreasing costs. Economies of scale are measured by comparing long-term marginal costs with long-term average costs. The maximum economies of scale that were assumed are: 0.5 for rail, 0.2 and 0.5 for capital and operating costs for Underground, 0.7 for buses and coaches, and 0.9 for cars. As far as possible, estimates of internal costs are based on incremental costs associated with greater use of that mode. We have assumed there is no variation in marginal external costs with changes in mode use, except in the case of congestion which is modelled explicitly as explained above.

Table 4.1 summarizes, for each mode in each of the three models (inter-urban, London peak and London off-peak) the estimates of marginal external costs (MEC) from each of the five main sources, and the long-term marginal internal costs (LRMC) of that mode (on the extreme assumptions of constant returns to scale, CRS, and maximum increasing returns to scale, EOS). Summation of the relevant long-term marginal cost with the estimated marginal external costs gives us an estimate of the efficient inclusive prices which is compared with, in the final column, our estimate of the current prices faced by users of each mode. The current price is based on existing prices and taxes, the efficient prices assume removal of all existing taxes and the imposition of direct charging for all identified external costs. All values are given in terms of pence/passenger-km at 1991 values based on estimates of average vehicle occupancy for each mode.

These figures suggest three basic findings. First, there are some important differences between efficient prices in the three situations. Inter-urban prices are lower than in the other two situations, but current prices tend to be higher than efficient prices, assuming maximum scale economies, whereas they are more likely to underestimate efficient prices in London. Second, given that scale economies are clearly very important for rail-based modes where external costs are relatively low, correct estimation of these scale economies becomes critical. Third, not unexpectedly, the external costs of car in urban areas dominate the internal costs, but do not do so in inter-urban travel. Rather surprisingly, however, the external costs of bus reinforce high internal costs and relatively small external economies in urban travel. This arises largely from recent evidence on particulate emissions which has increased dramatically the local air pollution costs of (mainly diesel) buses. This external cost is reinforced

Table 4.1. Marginal internal and external costs and prices of passenger transport (pence/passenger-km)[1].

Pence/ passenger-km	Global warming	Air pollution	Noise pollution	Congestion	Accidents	Total MEC	LRMC[2]	Efficient price with CRS[2,3]	Efficient price with EOS[3,4]	Current price
Inter-urban										
Rail	0.01	0.12	0.02	0.04	0.03	0.22	9.67	9.89	5.05	7.11
Car	0.02	0.35	0.08	0.85	0.15	1.45	5.15	6.60	6.08	7.78
Coach	0.01	0.39	0.01	0.15	0.01	0.57	3.00	3.57	2.67	3.09
London										
Underground										
Peak	0.01	0.13	0.09	0.72	0.03	0.98	45.18	46.16	10.02	10.12
Off-peak	0.01	0.13	0.09	0.00	0.03	0.26	15.80	16.06	8.16	8.94
Rail										
Peak	0.01	0.13	0.09	0.80	0.03	1.06	20.11	21.17	11.12	6.88
Off-peak	0.01	0.13	0.09	0.07	0.03	0.32	12.55	12.87	6.59	6.88
Car										
Peak	0.03	4.34	0.39	15.06	1.50	1.32	7.12	28.44	7.73	11.28
Off-peak	0.02	2.89	0.39	1.65	1.50	6.45	6.54	12.99	12.34	10.04
Bus										
Peak	0.01	7.21	0.09	3.73	0.88	11.92	15.27	27.19	22.61	10.63
Off-peak	0.01	5.41	0.09	1.76	0.88	8.15	13.00	21.15	17.25	10.63

[1] GB£0.01 = US$0.015 = ECU0.012.
[2] Assuming constant returns to scale.
[3] Efficient price is defined as LRMC + MEC.
[4] Assuming maximum returns to scale (see text for definition).

by rather poor revealed load factors found on local buses in the UK since deregulation, a finding confirmed by the work of Acutt and Dodgson (1996). Note that, in inter-urban travel, coach has much lower marginal internal and external costs.

In inter-urban travel, we have had to consider carefully the question of congestion for car external costs. These costs, although marginal in the sense of reflecting the cost to the marginal user, relate to a propensity to meet congestion averaged over the whole network. We have already noted the difficulties in modelling congestion on the inter-urban motorway network. We have made some allowance for this, but still require better data on the incidence of congestion to make a full correction. Recent work by Newbery (1994) and McLaren and Higman (1993) presents evidence suggesting that on certain motorways the marginal car and coach congestion costs are greater than the base estimates we have used from existing data. For peak travel on the most congested motorways, we estimate this would increase car and coach marginal congestion costs to around 2–3 pence/passenger-km and 0.5 pence/passenger-km, respectively. The figures in Table 4.1 should, therefore, be taken as a base estimate. We believe that our figures are the most consistent estimates, although such adjustment does not make a substantial difference to our final results. On congested urban routes the actual costs are much higher, as found for London where the peak period full cost of car usage is as high as 28 pence/passenger-km and 12–13 pence in the off-peak.

The London figures emphasize the importance of scale economies, especially for rail-based systems, in an urban situation. If constant returns to scale were in operation, it would be difficult to justify the Underground, but with the assumption of maximum scale economies the current prices appear to be very close to the estimated efficient prices. Rail is currently priced about right in the off-peak, but rather underpriced in the peak. Bus and car are both underpriced in both off-peak and peak periods.

4.4. Results

Using the evidence on demand, internal and external costs assembled above, the model shown in Figure 4.1 was constructed. For each of the differing assumptions about the degree of economies of scale, the consequences are examined for the demand for each mode, the degree of deficit financing assuming long-term marginal cost pricing, the total value of external effects, the implicit 'tax revenue' and the need for investment in additional capacity. Table 4.2 summarizes the results for demand changes in both short term (capacity fixed) and long term (assumed to be a period of 10 years) according to the degree of scale economies.

Given constant returns to scale, there is a tendency towards a short-term reduction in demand. However, in situations where congestion is less serious,

Table 4.2. Demand changes[a] using long run marginal cost pricing.

	Constant returns to scale				Increasing returns to scale			
	Short-term peak	Short-term off-peak	Long-term peak	Long-term off-peak	Short-term peak	Short-term off-peak	Long-term peak	Long-term off-peak
London								
Underground	−11.5	−6.8	−8.0	**15.7**	**6.0**	**10.9**	**17.5**	**24.8**
Rail car	−10.4	−12.8	**1.1**	**41.2**	−7.6	**11.9**	**10.7**	**57.4**
Bus	**0.8**	**1.3**	**28.8**	**40.9**	−4.0	−4.6	**19.7**	**31.1**
	−6.2	−12.5	−10.7	−8.6	−10.9	−14.9	−14.8	−15.3
	Short term		Long term		Short term		Long term	
Inter-urban								
Rail	−10.9		**18.5**		**9.2**		**48.3**	
Car	**7.1**		**48.1**		**2.9**		**42.4**	
Coach	**0.7**		**5.5**		−2.1		**2.6**	

[a]Changes given as percentage change from current demand.

or in the long term when the effects of a strong positive income effect are significant, there is a tendency for relative mode switching to car and an overall growth in the demand for all forms of transport. In the long term, growth in demand is driven by the income effect. Efficient taxation of externalities does not lead to greater demand for, or a relative shift to, modes of transport with lower external costs. With increasing returns to scale, however, the shift to car is less marked and there is overall growth of public transport modes, with the exception of the urban bus.

There are broad similarities between the London and inter-urban results. The London results suggest how the effects of congestion lead to a shift between the peak and off-peak. On the rail modes, where there are more likely to be capacity constraints, this is more pronounced. In this model the bus does not appear to perform well as a substitute for a car, due to its apparently poor environmental performance, the problems of road congestion and the lack of such substantial scale economies as on rail modes. This highlights the potential importance of technological improvement and bus priority schemes, which we have not allowed for in the model. The implementation of current moves towards 'greener' diesel engines or electric-powered vehicles in dense urban areas would lead to a reduction in local air pollution costs which would change the long-term results. Any use of road space for bus priority measures would have differential effects on bus and car congestion in the longer term.

In the inter-urban market, scale economies have a substantial impact on the long-term growth of rail demand. Given the relative sizes of car and rail markets, however, this does not result in a great shift from car to rail. If

Table 4.3. Financial implications of long-term demand estimates (£bn).

	New infrastructure capital requirements	Total annual price revenue (excluding tax)	Annual tax revenue	Total annual internal costs	Annual surplus/deficit
London – long-term Peak					
Underground	2.33	0.44	0.05	1.64	−1.20
Rail	1.07	0.61	0.08	1.23	−0.62
Car	4.58	1.62	4.56	3.97	−2.35
Bus	n.a	0.11	0.14	0.26	−0.15
London – Long-term off-peak					
Underground	n.a	0.28	0.01	1.04	−0.76
Rail	n.a	0.19	0.01	0.38	−0.19
Car	n.a	2.45	2.80	3.31	−0.86
Bus	n.a	0.24	0.25	0.49	−0.20
Inter-urban – Long term					
Rail	2.62	0.87	0.07	1.73	−0.86
Car	5.88	5.40	1.70	7.81	−2.41
Coach	0.01	0.60	0.16	0.90	−0.30

congestion on key links of the motorway system were to be modelled in more detail, or the decision taken not to implement any of the implied optimal additions to capacity, there would be a more pronounced shift from car to rail.

It is frequently argued that shifts in demand towards less environmentally damaging modes does not take place because of the perceived poor quality of service. A premise of this research was that the implementation of efficient pricing might not be possible because of supply (capacity) constraints in those modes. Data on service quality elasticities are still rather poor, but the model employed did enable an assessment of the likely need for new capital investment. Furthermore, it enabled a calculation of the implicit revenue from the optimal taxes/charges and the effects on the surplus/deficits of operators from the estimated demand changes. Table 4.3 gives the results for the long term, since this allows for any capital investment, and assuming significant scale economies. This implies the biggest impact on public transport, both in terms of its advantages and the likely financial impacts on operators from charging at a marginal cost well below average cost. The figures for the new investment in capital required are given as a total for the whole 10-year period, while the other figures are all given on an annual basis. Estimates of price revenue (excluding taxes) are based on prices equal to long-term marginal private costs, figures for the estimated annual tax revenue are those required to cover the marginal external costs. Comparing the last two columns of Table 4.3 enables an assessment of the potential for a self-financing transport system to be made. This shows that the inter-urban market, as defined here, requires substantial capital investment, but is not self-financing, with an annual shortfall of £1.6

billion. In London, in both the peak and off-peak, the tax revenue is more than sufficient to cover the deficits incurred as a result of long term marginal cost pricing. This generates an estimated surplus, of tax revenue over operating deficit, of about £1.5 billion/year. For presentation purposes, the level of investment in new infrastructure required is allocated entirely to the peak period and amounts to an annual average of about £0.8 billion. Treating London as self-contained would enable this investment to be financed from transport taxes. Alternatively, there is an obvious conclusion of the need to consider a national transport system where the excess tax revenues in urban areas are used to finance the deficits in inter-urban travel. This leaves the transport sector as potentially self financing, but not as a net contributor to general tax revenue.

4.5. Conclusions

We have presented a simple framework for analysing the effects of imposing efficient pricing on both inter-urban passenger travel in the UK and passenger travel in London. The models developed use estimates of the long-term marginal internal costs of the main modes, to which have been added estimates of marginal external costs from various sources in order to obtain the true opportunity cost of using each mode. Following removal of all existing subsidies and current taxation from the current prices, these could then be compared with the efficient prices in order to derive predictions of the changes in demand for each mode, together with implied tax revenues and operators' aggregate surplus or deficit.

The objective of this research has not been to provide a detailed forecast of what will happen, but rather to develop a modelling framework which can help in the analysis of what might result from the adoption of various policies. We are conscious that there may be continuing controversy over some of the values of the external effects used here (see, for example, the higher values on global warming used by Mauch et al., 1995); we believe we have used correct values, but the model easily allows for use of different values. We have not reported the results here, but substantial sensitivity testing of differing values of global warming effects, for example, suggests no major difference in the overall results. The results are also highly dependent on the assumed values of elasticities. We also recognize that the values of such elasticities may be less appropriate in an age in which there are strong exhortations to changes in travel behaviour, and where also there may be important shifts in technology. This framework will allow testing of any such improved information when it becomes available.

The approach adopted here has two important advantages over most attempts to develop efficient pricing models: first a long-term approach which explicitly allows for investment in capacity is required to meet the constraints which may otherwise prevent modal shift. Second, it has explored the important role of scale economies in justifying a shift to environmentally less damaging modes of transport. The investment is costed in terms of the actual costs of new investment for each mode, a factor which is open to policy driven variation (e.g. through decisions on land use, protection of natural environments etc). Scale economies remain, however, a source of difficulty in that most operators are relatively ignorant of the true level of scale economies present or potentially available.

The important conclusion to be drawn from this paper is that efficient pricing and taxation of externalities is not sufficient to give substantial shifts to modes of transport with lower external costs. De Borger et al. (1996) have found similar results from a passenger transport model of urban areas in Belgium. Their model does not, however, consider scale economies, the costs of infrastructure provision or long-term demand. In the long term, demand changes are driven mainly by the growth of income. The assumption made about the appropriate scale economies present is an important determinant of the degree of change and overall effects. The present levels of taxation of car users are such that the current prices of inter-urban road transport are close to (perhaps slightly higher than) the efficient level. In London, especially in the peak, car transport is currently underpriced by a substantial amount, although so are bus and rail transport (albeit to a lesser extent), with the exception of the underground. The deficits incurred by public transport operators in moving to long-term marginal costs pricing including all external effects could be financed from the revenue raised by efficient taxation of car travel. Taking London on its own there is a clear surplus of revenue over the direct deficits, but this is not so in the case of inter-urban travel and the overall position is not clear cut. After financing such deficits, the transport sector would not be a major net contributor to government revenue.

The conclusion that, in the short term, efficient pricing does not result in large shifts in demand to the modes of transport with lower external costs has important implications for transport policy (compare, for example, Commission of the European Communities, 1995). Since it seems unlikely that a market solution can lead to substantial reductions in traffic growth or falls in external costs, it may be necessary to educate consumers into a change in life-styles advocated by the recent Royal Commission on Environmental Pollution (1994) report. This interpretation has important implications for transport policies that are based on the market mechanism.

This model is only a partial one of the transport sector. It does not include the interaction between freight and passenger traffic using the same infrastructure. It also fails to allow for the fact that most inter-urban journeys by road involve urban segments at each end of the journey, such that their overall cost is higher than implied here. There is an interaction between the London and inter-urban models which would be feasible, but complex, to incorporate. What has been shown here is that it is possible to shed some empirical light on the questions raised in current debates, but also that transport policy may need to be much more pro-active in setting clear goals as to appropriate traffic levels and modal balance if it is to achieve a lasting change in use. How those goals should be set independently of the efficient pricing and concepts of optimality used here is beyond the scope of this chapter, but the model developed here can be used to assess the effects of such policies through their impacts on key values, elasticities or investment rules.

Notes

1. This paper is based on a project financed under the Transport and the Environment Programme of the UK Economic and Social Research Council (Grant L119251009). The research assistance of Ian Skinner is gratefully acknowledged. The valuable comments of participants in the Workshop on Environment and Transport in Economic Modelling, Venice, November 1995, have helped to improve the argument in this paper.

References

Acutt, M.Z. and J.S. Dodgson, 1996, Transport and global warming, paper to ESRC Conference on Transport and the Environment, London.

British Railways Board, 1993, Annual Report and Accounts 1992–93, London.

Button, K.J., 1993, Transport Economics, Aldershsot: Edward Elgar.

Commission of the European Communities, 1995, Towards Fair and Efficient Pricing in Transport: Policy Options for Internalising the External Costs of Transport in the European Union, Brussels.

De Borger, B., I. Mayeres, S. Proost and S. Wouters, 1996, Optimal pricing of urban passenger transport: a simulation exercise for Belgium, Journal of Transport Economics and Policy, 30, 31–54.

Department of Transport, 1981 and revisions, COBA 9, London: Department of Transport.

Department of Transport, 1991, Great Britain Transport Statistics 1991, London: HMSO.

Department of Transport, 1992, Great Britain Transport Statistics 1992, London: HMSO.

Department of Transport, 1993a, Bus and Coach Statistics Great Britain 1993, London: HMSO.

Department of Transport, 1993b, Road Traffic Statistics Great Britain 1993, London: HMSO.

Department of Transport, 1994a, Trunk Roads and the Generation of Traffic, Standing Advisory Committee on Trunk Road Assessment (SACTRA), London: HMSO.

Department of Transport, 1994b, The Government's Response to the SACTRA Report, London: HMSO.

McLaren, D.P. and R. Higman, 1993, The Environmental Implications of Congestion on the Interurban Network in UK, paper given to 21st PTRC SAM 1993, Seminar F.

Maddison, D., O. Johansson, D. Pearce, E. Calthrop, T. Litman, and E. Verhoef, 1996, Blueprint 5: The True Cost of Road Transport, London: Earthscan.

Mauch, S.P., W. Rothengatter et al., 1995, External Effects of Transport Report by IWW, Karlsruhe and INFRAS, Zürich, Paris: International Union of Railways (UIC).

Newbery, D.M., 1987, Road User Charges in Britain, Centre for Economic Policy Research, Discussion Paper No. 174.

Newbery, D.M., 1988, Road user charges in Britain, Economic Journal, 98, 161–176.

Newbery, D.M., 1990, Pricing and congestion: economic principles relevant to road pricing, Oxford Review of Economic Policy, 6, 22–38.

Newbery, D.M., 1994, The case for a public road authority, Journal of Transport Economics and Policy, 28, 235–253.

Pearce, D.W., 1993, Blueprint 3: Measuring Sustainable Development, London: Earthscan.

Peirson J., I. Skinner and R. Vickerman, 1994a, Environmentally efficient transport taxes and investment, CERTE Discussion Paper 94/1, Canterbury: University of Kent.

Peirson J., I. Skinner and R. Vickerman, 1994b, The taxation of external effects and the efficient supply of transport, CERTE Discussion Paper 94/3, Canterbury: University of Kent.

Peirson, J., I. Skinner and R. Vickerman, 1995, Estimating the external costs of UK passenger transport: the first step towards an efficient transport market, Environment and Planning A, 27, 1977–1993.

Peirson, J. and R. Vickerman, 1996, Efficient pricing and optimal investment in transport infrastructure, CERTE Discussion Paper 96/1, Canterbury: University of Kent.

Peirson, J. and R. Vickerman, 1997, Environmental effects of transport: a model of optimal pricing and investment for the UK, International Journal of Environment and Pollution, 7, 343–356.

Peirson, J. and R. Vickerman, 1998, The environment, efficient pricing and investment in transport: a model and some results for the UK, in D. Banister (ed.) Transport Policy and the Environment, London: Chapman and Hall.

Royal Commission on Environmental Pollution, 1994, Transport and the Environment, Eighteenth Report, Cm 2674, London: HMSO.

CHAPTER 5

Carbon Emissions and the Economic Costs of Transport Policy in Sweden[1]

Glenn W. Harrison and Bengt Kriström

In recent years the Swedish transport system has become a target for intensive political discussion. One can single out the environment and infrastructure as typical buzzwords of this debate. Sweden has invested a significant amount of its prestige in showing that it can stick to agreements made in conjunction with the Rio Summit in 1992. The growing transport sector is a key challenge, and perhaps it is here that Sweden will face the most substantial difficulties in meeting the obligations. Nevertheless, the notion that it is possible to create an 'environmentally friendly' transport sector has become a theme of many recent proposals on the future of Swedish transport policy. We evaluate several of these proposals by constructing and simulating a computable general equilibrium (CGE) model of Sweden.

5.1. The transport policy debate in Sweden

5.1.1. Current issues

Sweden's transport policy is based on five objectives, as indicated in the 1988 Transport Policy Resolution: availability, efficiency, safety, environmental quality and regional balance. A substantial number of reports have assessed the success of the policy during the past few years. According to a recent assessment, (Statens Offentliga Utredningar, 1996), a number of improvements have been secured. For example, safety has improved; the target set previously of a maximum of 600 fatalities per year has been met. Deregulation of air traffic and the introduction of new high-speed trains have contributed to the efficiency of the transport system, although a complete evaluation of airline deregulation remains to be undertaken. Emissions of certain pollutants have diminished considerably, although CO_2 emissions from the transport sector have increased.

R. Roson and K.A. Small (eds.), Environment and Transport in Economic Modelling, 76–117.

A number of different proposals to mitigate those emissions have been put forward, mainly from various recent government Commissions and quasi-government Committees. We return to these proposals and the evolving debate around them below.

The 1988 Transport Policy Resolution suggested a number of guiding principles for costing the transport system. An important principle is that charges are to be set in proportion to social costs. Indeed, Sweden has since 1988 introduced environmental taxes on SO_2, NO_2 and CO_2 emissions. Whether or not the levels of those taxes have been set according to the marginal cost of damage from the relevant externality is impossible for us to say. Still, a general contention is that these taxes have led to reductions in emissions, *ceteris paribus*. A detailed assessment by the so-called Green Tax Commission was published in 1997 (see Harrison and Kriström, 1997).

One of the key ingredients in the 1988 and 1991 Environmental Policy Resolutions is the principle of sector responsibility. Thus, rather than having general environmental goals for the whole economy, goals should be broken down on the sector level. Since then, a number of goals have been suggested for the transport sector, sometimes detailing particular types of traffic. For example, the Air Aviation Board has announced a target to stabilize 2010 emissions of CO_2 from air traffic to the 1990 level. This is different from the national goal of stabilizing to 1990 levels by the year 2000.

Recent proposals to reduce the emissions of greenhouse gases have originated from individual political parties, from non-governmental organizations and from various official investigations. We discuss two of those here.

The main task of the Traffic and Climate Committee (Statens Offentliga Utredningar, 1995), was to propose measures for reducing emissions of carbon dioxide and other greenhouse gases from the transport sector. Because Sweden is prevented by international agreements from taxing fuels for air traffic and shipping in the same way as for road transport, the proposals focused on measures to reduce road traffic emissions. The committee concluded that CO_2 emissions from the transport sector should not increase up until the year 2005. Again, this is different from the overall environmental goal for Sweden, but need not be inconsistent with it. The price of petrol should be raised by SEK 0.40/litre with effect from January 1, 1997 and for 4 years subsequently. The tax increase should hit all fossil fuels and be uniform across all sectors. In mid-1996 the price of petrol in Sweden was about 8 SEK/l (roughly US$1.23/l). Because the carbon tax today is 0.37 SEK/l, the proposal effectively means a doubling of the CO_2 tax on petrol over 4 years.

The committee pointed out that the CO_2 target was not independent of the development of Sweden's energy policy. A key issue here is the destiny of nuclear power, currently planned to be decommissioned by 2010. The Commission argued that the CO_2 target should be reassessed in conjunction with the development of future energy policy and the development of international agreements on greenhouse gases (see Nordhaus, 1995, for further discussion).

The so-called 'KomKom' Commission (Commission on Communications; Statens Offentliga Utredningar, 1996) proposed that the CO_2 tax be increased such that the real price of petrol increases by 0.10 SEK/year between 1998 and 2020. The same increase is proposed for diesel. In this way, the petrol price would be increased to SEK 2.30/l by 2020. The tax revenues should be returned to the transport sector in the form of government support of environmental measures. The Commission argued that this proposal would have substantial, indeed unacceptable, distributional effects in certain regions of the country and suggested some regional policy measures to lessen the regressive impacts of the proposals.

5.1.2. *The general structure of Swedish energy taxes*

Sweden has used taxes on energy since 1929, when a tax on petrol was introduced. Electricity has been taxed since 1951, followed by a broadening of the energy taxes in 1957. The motivation underlying these taxes was purely financial. In the 1970s, propelled by the global energy crisis, energy taxes were increasingly motivated by a desire to discourage consumption of fossil fuels. Thus, increased taxes on oil products were coupled by a significant expansion of electricity supply in order to promote a different profile of energy consumption.

Environmental concerns entered the discussion in the 1980s, manifested by the introduction of a tax differentiation of leaded petrol in 1986. This was followed by the Environmental Tax Commission that recommended a rich array of environmental taxes in their final proposal (see Statens Offentliga Utredningar, 1990: 59), This investigation led the government to propose taxes on emissions of CO_2 and sulphur, *inter alia*, in 1991. While this was not the first official body in Sweden to discuss environmental taxes, this mission was unique in that it was coupled with a major overhaul of the Swedish tax system in the beginning of the 1990s. The general tax reform included a reduction of income taxes, to be financed partially by an increased use of energy and environmental taxes (including the introduction of VAT on energy consumption).[2]

For the purpose of harmonizing Swedish energy taxes with those prevalent within the most important competing countries, another reform of energy taxation passed on January 1, 1993. This reform was closely tied with the international competitiveness concerns that have been a recurring issue in the design of Swedish energy policy. It meant that manufacturing industry no longer paid energy tax on the use of fuels and electricity in their processes. In addition, there was a reduction in the CO_2 tax for the manufacturing industry, as detailed below.

Industry exemptions

In an international context Swedish energy taxes are high. Because export-oriented industries are competing on markets with significant price elasticities, it is not surprising that several tax exemptions are available. Beginning in 1974, through the law on (partial) exemptions of the general energy tax, energy-intensive manufacturing industries and the horticulture industry have escaped some part of energy taxes. This, of course, is not unique in Europe. Similar exemptions have also been used in Denmark and Norway for manufacturing.

These exemptions for manufacturing are a key feature of the tax system we evaluate. In the tax system prior to 1993 approximately 100 energy-intensive firms were granted reduced tax rates on fuels and electricity. In 1992 the reduction for energy-intensive industry was worth 1.3 billion SEK. The new energy and carbon tax system introduced in 1993 resulted in significantly reduced tax rates for industry. The total amount of energy and carbon tax collections dropped from 3.8 billion SEK in 1992 to just 0.5 billion SEK in 1994. We approximate these exemptions as applying to manufacturing industries *in toto*, so that manufacturing industry and horticulture are assumed to pay 25% of the general carbon tax rate.

Before the 1993 change of the energy tax system, tax exemptions were essentially granted on a case-by-case basis. Thus energy-intensive industries could apply for a reduction of the energy tax on electricity and fuels. With a zero energy tax on electricity and fossil fuels, such applications are now redundant. It is still possible to obtain deductions for fuel use, some of which are of considerable importance for individual firms (see Statens Offentliga Utredningar, 1994). These deductions are only available to firms producing cement, lignite and glass. They only apply to the carbon tax on coal and natural gas, and not on the use of oil products. In 1995 less than 10 energy-intensive firms benefited from this rule, and the value of the reduced tax was less than 50 million SEK.

The CO_2 tax

By far the most important of the environmental taxes introduced as the result of the Environmental Tax Commission is the CO_2 tax. Introduced in January 1991, the tax of 0.25 SEK/kg emitted CO_2 was followed by intense controversy. Eventually, a reform of energy taxes in 1993 led to significant reductions for manufacturing industries, as explained above. The government argued that it was important to reduce Swedish energy taxes to European levels for internationally competitive industries, lest firms move abroad or remain at a significant cost disadvantage. Carbon taxes in Sweden in 1995, the base year of the model's representation of the tax system, are generally about 0.34 SEK/kg emitted CO_2 for non-exempted sectors and 0.083 SEK/kg for manufacturing sectors.

5.1.4. *The European Union*

An advisory referendum held in Sweden in November 1994 resulted in a 52% to 47% win for the proponents of entering the EU. As a result Sweden has been a member of the EU since January 1995. It is not currently clear what

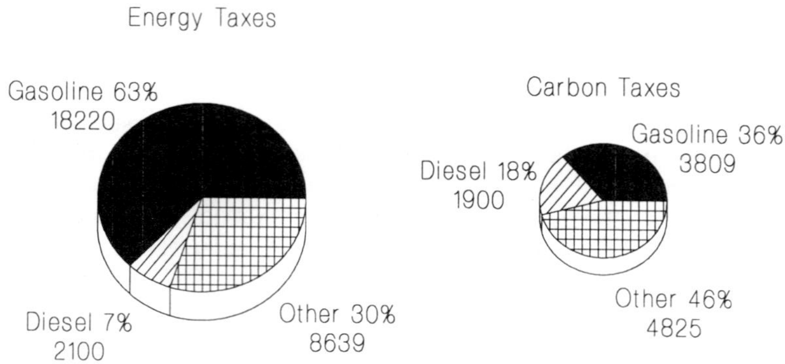

Figure 1. Revenues from energy taxes in 1994 (millions of SEK).

kinds of restrictions there will be on the possibilities of pursuing an independent environmental policy. On the one hand, current EU policy is based on minimum requirements, which means that a member country has an option to use a stricter policy. On the other hand, it is difficult to block imports of goods that have been approved in another country. Membership in the EU does not prevent country-specific environmental policies *de jure*, but it may make a deviation from EU policy impossible *de facto*.

When Sweden entered the EU a new energy tax law (Statens Offentliga Utredningar, 1994) replaced the old laws on general energy taxes, CO_2 taxes, sulphur taxes, petrol taxes and diesel taxes. The new law substantially harmonizes Swedish rules with those in the EU. Generally, the above taxes are due on fuels used for heating purposes, or as propellants for engines. Biofuels are exempted from energy taxes, following a long tradition in Swedish energy policy to encourage substitution towards these fuels. Fossil fuels and electricity used in manufacturing are treated favourably, the motivation again being the concern with international competitiveness.

Current Swedish energy taxes generated about 40 billion SEK in 1994. The structure of these revenues, in terms of the CO_2 tax and other energy taxes, is shown in Figure 5.1. The total revenues from energy and environmental taxes in 1994, including sales taxes on motor vehicles and annual road taxes, were roughly 47 billion SEK (Treasury of Sweden, 1995; p. 60). This corresponds to about 6% of total tax revenues (Treasury of Sweden, 1995; Fig. 13.1) or about 3% of GDP.

5.2. A general equilibrium model

5.2.1. Basic features

Our small open economy (SOE) model is designed for tax policy analysis with a large number of sectors. The model is a generic general equilibrium model of a single economy along the lines of Melo and Tarr (1992), Harrison et al. (1993a) and Rutherford et al. (1994). We describe here the general features of the base model, adding details about the 1992 version for Sweden later. Further details on the database construction are provided in Harrison and Kriström (1997, Appendix A). The complete database and model is available in machine-readable form from web page http://theweb.badm.sc.edu/glenn/sweden.htm.

Goods are produced using primary factors and intermediate inputs. Primary factors include capital and six types of labour. Production exhibits constant returns to scale and individual firms behave competitively, selecting output levels such that marginal cost at those output levels equals the given market price. Output is differentiated into goods destined for the domestic and export markets. Exports are further distinguished according to whether they are destined for specific foreign markets. This relationship is characterized by a two-level constant elasticity of transformation frontier. Composite output is an aggregate of domestic output and composite exports; composite exports are aggregates of exports for distinct foreign markets.

Final demand by private households arises from nested constant elasticity of substitution (CES) utility functions. This allows consumer decision-making to occur in the form of multi-stage budgeting. At the top level the consumer trades off a composite bundle of consumer goods with leisure (the own-consumption of the consumer's labour endowment). At the second level goods from different sectors compete subject to the budget constraint of the consumer, and all income elasticities are unity. In the third stage the consumer decides how much to spend on domestic or imported goods in each sector, subject to income allocated to spending in that sector in the first stage. Finally, having decided how much to spend on imports as a whole, the consumer allocates this expenditure on imports from specific countries. Each allocation decision is modelled as a CES function.

The model allows tariff rates to differ depending on whether the imports are from specific trading partners. Exports can be sold at different prices depending on whether they are destined for distinct foreign markets. The same is possible on the import side.

Government expenditures and investment demand are exogenous. Funding of government expenditures is provided by tax revenues and tariff revenues. In addition to tariffs, the government also derives income from indirect taxes (net of subsidies). These are modelled as value added taxes (VAT). Unless otherwise specified the government recovers any lost revenues by increasing taxes on labour collected at the enterprise level; similarly, it reduces those taxes for any increase in revenue due to a counter-factual scenario.

Since private consumption equals the income from primary factors plus net transfers to the consumer by the government (from domestic and foreign trade taxes), Walras law is satisfied. Changes in public consumption are balanced with changes in revenue, so that the public deficit in the base year is effectively exogenous.

World market import and export prices are fixed, so there are no endogenous changes in the terms of trade. In other words, import supplies and export demands are infinitely elastic at given world prices. The current account imbalance in the base year is assumed to be matched by an exogenous capital inflow or outflow. These capital flows have no affect on the stock of domestic capital, nor on interest payments to foreigners. Domestic prices change to ensure that the change in the current account is zero. The fixed world prices that Sweden is assumed to face may be changed parametrically.

5.2.2. *The Swedish model*

Based on 1992 input–output data for Sweden, the model identifies 87 sectors.[3] These are listed in Table 5.1, along with their pseudo-Swedish acronym. This is the level of disaggregation available through the input–output statistics, and provides excellent detail for our purposes. It is possible to aggregate to a smaller number of sectors, such as has been popular in previous CGE models of Sweden, but there seems little advantage in doing so and potential for misleading analysis in the present context.[4] Moreover, it is always possible to assess the information loss of employing specific aggregations if the model is fully disaggregated, while the reverse is obviously not true.

The household disaggregation is based on the 1992 Household Expenditure Survey conducted by the Statistiska Centralbyrån (SCB), which provides detailed information on expenditure patterns of 30 households. These households are differentiated by family status and income, and are listed along with their acronyms in Table 5.2. One difficulty is that the expenditures of each household are defined over consumer goods, and no ready mapping exists from our industrial products to those goods. We resolve this problem by using our intuition, and using the data from the household expenditure survey to allow different households to have different expenditure patterns for different industrial goods.

We also assume that each household receives its income from slightly different sources. In other words, each household has a slightly different share of each primary factors in its endowment. In the absence of better data, we are not overly confident of this feature of the model, and prefer to view households as being primarily distinguished on the basis of their expenditure patterns. Hence we primarily capture variations in the cost of living for different households, and probably do not capture all of the variations in the value of endowment income for different households.

Primary factors are used in the production of value added in each sector. In general two types of factors are free to move across sectors to equate after-tax rates of return: labour and capital (K). Labour is differentiated by skill categories and occupational status into six groups: blue collar unskilled (L_BC_U), blue collar skilled (L_BC_S), white collar unskilled (L_WC_U),

Table 5.1. Sectors in the Swedish model.

JORD	JORDBRUK	Agriculture and hunting
SKOG	SKOGSBRUK	Forestry and logging
FSKE	FISKE	Fishing
JARN	JARNGRUVOR	Iron ore mining
A_ME	A MET.GRUVOR	Other metal mining
STEN	STENBROTT A.GR.	Stone quarrying and other non-metallic mining
SLAK	SLAKTERIER	Meat slaughtering
MEJE	MEJERIER	Dairy products
FRUK	FRUKTKONSERVER	Canning of fruits and vegetables
FISK	FISKKONSERVER	Canning of fish
FETT	FETT OLJOR	Oils and fats
KVAR	KVARNPRODUKTER	Grain mill production
BAGE	BAGERIPROD.	Bakery products
SOCK	SOCKER	Sugar
CHOK	CHOKLAD KONF.	Confectionary
DIVX	DIV.LIVSMEDEL	Other food
FODE	FODERMEDEL	Prepared animal feeds
DRYC	DRYCKER	Beverages
TOBA	TOBAK	Tobacco
GARN	GARN VAVNAD	Spinning and weaving
TEXT	TEXTILSOMN.	Textiles other than clothing
TRIK	TRIK2VAROR	Hosiery and knitted goods
OVRT	OVR TEXTIL	Other textiles
BEKL	BEKLADNAD	Clothing
LADE	LADER SKOR	Leather and shoes
S2GV	S2GVERK	Wood preparations
TRAH	TRAHUS SNICK.	Wooden building materials
A_TR	A TRAMATERIAL	Other wooden materials
OVR_	OVR TRAVAROR	Other wood products
TRAM	TRAMOBLER	Wooden furniture
PAPP	PAPPERSMASSA	Paper pulp
PPPP	PAPPER PAPP	Paper and board manufacturing
TRAF	TRAFIBERPL.	Fibreboard
PFRP	PAPPFORP.	Paper packaging products
OVRX	OVR. PAPPER	Other paper products
GRAF	GRAFISK IND	Printing and publishing
KEMI	KEMIKALIER	General chemicals
GODS	GODSELMEDEL	Fertilizers and pesticides
BASP	BASPLAST	Plastics and synthetic fibres
PLAS	PLAST HALVF.	Semi-finished plastic products
FARG	FARG	Paints
LAKE	LAKEMEDEL	Drugs and medicines
TVAT	TVATTMEDEL	Soaps and detergents
OVRK	OVR KEMIK.	Other chemical products
PETR	PETROL.RAFF	Petroleum refining
SMOR	SMORJMEDEL	Lubricating oils and greases
GUMM	GUMMIVAROR	Rubber products
PLSV	PLASTVAROR	Plastic products
PORS	PORSLIN	Pottery

Continued.

Table 5.1. (Continued.)

GLAS	GLAS	Glass and glass products
TEGE	TEGEL	Structural clay products
CEME	CEMENT	Cement and plaster
OVRM	OVR MINERAL	Other non-metallic mineral products
JRN_	JRN O ST2L	Iron and steel
FERR	FERROLEGERING	Ferro-alloys manufacturing
JNGJ	JNGJUTERIER	Iron and steel casting
META	METALLVERK	Metal fabrication
METV	METALLVALSV.	Metal rolling mills
I_JA	I JARNGJUTERI	Iron and steel casting
METR	METALLVAROR	Other metal casting
MSKN	MASKINER	Industrial machinery
ELMO	ELMOTORER	Electrical machinery
TELE	TELEPRODUKTER	Electronics and telecommunications
HUSH	HUSH2LLSMASK.	Domestic eletrical appliances
OVRE	OVR.ELPROD.	Other electrical goods
VARV	VARV B2TAR	Ship building and repair
RALS	RALSFORDON	Railroad building and repair
BILA	BILAR	Motor vehicles and parts
CYKL	CYKLAR	Bicycles and motorcycles
FLYG	FLYGPLAN	Aircraft manufacture and repair
OVRR	OVR TRANSP.M.	Other transport equipment
INST	INSTRUMENT	Scientific instruments
A_TI	A TILLVERKN.	Other manufacturing
EL_O	EL O VARMEVERK	Electricity and steam
GASV	GASVERK	Gas
VATT	VATTENVERK	Water
BYGG	BYGGNAD	Construction
VARU	VARUHANDEL	Trade
HOTE	HOTELL REST.	Hotels and restaurants
SAMF	SAMFARDSEL	Transport and storage
POST	POST TELE	Communication
BANK	BANK FORSAKR.	Banks and insurance
EGNA	EGNAHEM FRITID	Housing
FAST	FASTIGHETSFORVALTN	Other real estate
UPPD	UPPDRAGSV.	Business services
REPA	REPARATIONER	Repair services
OVRP	OVR. PR. TJ	Personal services

white collar semi-skilled (L_WC_SS), white collar skilled (L_WC_S) and self-employed (L_SE). The distribution of labour types in each sector is shown in Table 5.3. We allow the labour types to substitute with each other at a different rate than their composite does with K, although our formulation allows all primary factors to be equally substitutable as a special case.[5]

The model allows the specification of sector-specific capital types in any set of sectors. This possibility allows the identification of sectors that employ a significant amount of a primary factor that can be interpreted as specific to that sector. We could interpret this as referring to some short term in which

Table 5.2. Households in the Swedish model.

S_NC_1	Single adults with no children – first quartile
S_NC_2	Single adults with no children – second quartile
S_NC_3	Single adults with no children – third quartile
S_NC_4	Single adults with no children – fourth quartile
S_C_1	Single adults with children – bottom half
S_C_2	Single adults with children – top half
M_NC_1	Multiple adults with no children – first quartile
M_NC_2	Multiple adults with no children – second quartile
M_NC_3	Multiple adults with no children – third quartile
M_NC_4	Multiple adults with no children – fourth quartile
M_1C_1	Multiple adults with 1 child – first quartile
M_1C_2	Multiple adults with 1 child – second quartile
M_1C_3	Multiple adults with 1 child – third quartile
M_1C_4	Multiple adults with 1 child – fourth quartile
M_2C_1	Multiple adults with 2 children – first quartile
M_2C_2	Multiple adults with 2 children – second quartile
M_2C_3	Multiple adults with 2 children – third quartile
M_2C_4	Multiple adults with 2 children – fourth quartile
M_3C_1	Multiple adults with 3 or more children – first quartile
M_3C_2	Multiple adults with 3 or more children – second quartile
M_3C_3	Multiple adults with 3 or more children – third quartile
M_3C_4	Multiple adults with 3 or more children – fourth quartile
O_NC_1	Others with no children – first quartile
O_NC_2	Others with no children – second quartile
O_NC_3	Others with no children – third quartile
O_NC_4	Others with no children – fourth quartile
O_C_1	Others with children – first quartile
O_C_2	Others with children – second quartile
O_C_3	Others with children – third quartile
O_C_4	Others with children – fourth quartile

capital is applied to sectors in a manner that does not permit it to be readily moved to other sectors.[6] Instead, we use it to capture the limited range of activities to which resources can be applied. As one increases parametrically the assumed share of benchmark payments to K that is attributable to such specific factors, and thereby decrease the share that is assumed to attributable to the mobile K, the corresponding supply curve for that industry becomes more inelastic. The intuition is clear: as the relative demand for output for that industry falls, *ceteris paribus* all input prices, the factor that is specific to this industry cannot escape to other sectors. It must therefore experience a larger drop in real return than when it is inter-sectorally mobile and facing the same drop in derived demand for it's value marginal product. This relatively sharp decline in factor input cost results in a larger drop in the supply price

Table 5.3. Labour types in the Swedish model (percent employment in sector).

Sector	Blue collar			White collar					
	L_BC_U	L_BC_S	L_BC	L_WC_U	L_WC_SS	L_WC_S	L_WC	L_EMP	L_SE
JORD	18	11	29	6	4	6	16	46	54
SKOG	40	15	55	10	12	11	33	89	11
FSKE	25	3	28	4	5	56	65	92	8
JARN	35	30	65	6	17	8	31	97	3
A_ME	35	34	69	7	13	7	27	96	4
STEN	26	15	40	16	18	19	52	93	7
SLAK	41	25	66	11	8	9	28	94	6
MEJE	47	7	54	18	12	12	41	95	5
FRUK	41	7	48	17	13	17	47	95	5
FISK	56	4	60	11	8	13	33	93	7
FETT	32	13	45	13	18	19	50	95	5
KVAR	31	15	46	12	13	22	47	93	7
BAGE	40	23	63	14	6	10	30	93	7
SOCK	32	25	57	9	14	13	37	94	6
CHOK	48	8	56	15	11	13	39	94	6
DIVX	39	8	47	18	12	16	47	94	6
FODE	38	11	50	20	9	16	45	95	5
DRYC	43	8	52	15	13	14	41	93	7
TOBA	45	14	59	8	13	15	37	96	4
GARN	55	7	63	12	10	10	32	95	5
TEXT	55	8	64	12	6	12	31	94	6
TRIK	61	7	67	11	8	9	29	96	4
OVRT	42	10	51	11	12	20	43	95	5
BEKL	57	10	67	10	7	11	27	94	6
LADE	58	8	66	9	5	14	28	95	5
S2GV	60	13	73	8	6	8	22	96	4
TRAH	30	34	64	10	11	11	32	96	4
A_TR	56	16	72	8	9	8	24	96	4
OVR_	57	11	68	8	9	9	26	94	6
TRAM	42	27	69	8	8	11	26	95	5
PAPP	34	31	65	10	15	8	32	97	3
PPPP	40	21	62	11	14	10	35	97	3
TRAF	45	22	67	9	11	10	29	96	4
PFRP	37	17	55	13	14	14	41	96	4
OVRX	38	20	58	13	10	16	38	96	4
GRAF	14	29	43	17	22	12	50	93	7
KEMI	18	20	38	13	23	21	57	95	5
GODS	26	27	53	9	15	17	41	95	5
BASP	24	25	49	13	21	13	47	96	4
PLAS	43	13	56	11	12	16	40	95	5
FARG	31	6	37	19	19	20	58	94	6
LAKE	15	6	20	13	27	35	75	96	4
TVAT	33	4	37	25	13	20	57	94	6
OVRK	35	14	49	13	17	17	47	96	4
PETR	13	23	35	10	35	14	60	95	5

Continued.

Table 5.3. (*Continued.*)

Sector	Blue collar			White collar					
	L_BC_U	L_BC_S	L_BC	L_WC_U	L_WC_SS	L_WC_S	L_WC	L_EMP	L_SE
SMOR	31	10	40	17	20	18	54	94	6
GUMM	54	6	61	9	12	12	34	95	5
PLSV	49	12	61	10	11	12	33	94	6
PORS	48	13	61	10	12	13	35	96	4
GLAS	51	17	68	8	10	10	28	96	4
TEGE	42	14	56	15	12	13	40	96	4
CEME	32	22	54	12	14	16	42	96	4
OVRM	44	15	59	11	14	11	36	96	4
JRN_	40	23	63	9	15	10	34	97	3
FERR	43	24	67	10	9	9	28	95	5
JNGJ	45	24	68	7	11	8	26	94	6
META	42	24	67	9	12	8	29	96	4
METV	43	17	60	11	14	11	36	96	4
I_JA	47	23	70	6	11	8	25	95	5
METR	31	31	62	8	12	12	32	95	5
MSKN	16	31	48	10	21	17	48	95	5
ELMO	17	27	44	9	25	18	52	95	5
TELE	15	16	31	10	31	23	63	95	5
HUSH	33	28	60	9	14	11	34	94	6
OVRE	27	23	49	10	21	15	46	95	5
VARV	15	41	56	7	17	14	38	94	6
RALS	17	48	65	6	16	9	31	96	4
BILA	34	22	57	6	20	12	38	95	5
CYKL	44	13	57	13	12	14	39	96	4
FLYG	12	28	39	8	31	18	57	97	3
OVRR	41	20	61	10	11	15	35	96	4
INST	12	21	33	11	26	24	62	95	5
A_TI	33	23	56	13	9	16	38	94	6
EL_O	7	28	35	11	36	14	61	97	3
GASV	4	15	19	11	28	35	74	93	7
VATT	6	46	53	9	27	8	44	97	3
BYGG	19	11	30	10	29	21	60	90	10
VARU	26	8	35	26	11	21	59	94	6
HOTE	25	27	52	12	13	10	35	88	12
SAMF	40	9	49	17	11	15	44	93	7
POST	54	2	56	17	13	10	40	96	4
BANK	2	1	2	27	37	30	93	95	5
EGNA	26	15	41	19	19	15	53	94	6
FAST	5	3	8	19	26	38	83	91	9
UPPD	6	46	53	13	7	21	41	94	6
REPA	23	20	43	15	5	30	49	93	7
OVRP	26	15	40	16	18	19	52	93	7
Total	26	15	41	16	18	19	54	95	8

in that industry than when the factor is assumed mobile. The converse argument applies to increases in demand in the industry, of course.

Thus we can arbitrarily constrain the supply response of resource-based industries by specification of this parameter.[7] Given that the primary policy focus of these simulations is on the use of fossil fuels, such assumptions may be important. We assume that the share of observed payments to K that are payments to K that is specific to that sector is 0.2 for sectors STEN, PETR and SMOR, and 1 for sectors JORD, SKOG and FSKE.

Each sector produces output using intermediate inputs and a value added composite of the primary factors. Although the natural assumption[8] might be to model the substitutability of the intermediate inputs by assuming a Leontief technology,[9] we use instead a CES function with a low elasticity of substitution (0.25) across all sectors. This specification allows for later evaluation of the effects of varying degrees of substitutability at the point at which energy taxes typically impact in Sweden. The value added composite is produced using a CES production function and consists of two inputs: a labour composite and a capital composite. Each of these composites, in turn, is produced in a lower CES nest.

Trade is modelled as occurring at fixed world prices. However, Swedish importers may substitute between alternative import sources, and indeed between domestic production and an import composite. Similar assumptions apply on the export side, where Swedish producers have a constant elasticity of transformation between sales to domestic markets and a composite foreign market and sales of the composite export to any of several foreign trading partners. The key feature of our model in these regards is that Swedish producers have no market power in world markets.

In the present version we identify trade with Finland, Norway, Denmark, the rest of the EU, Japan, the United States, and a residual rest of world (ROW). Hence there are seven trading partners in the model. No data are available to allow us to identify different tariff rates or NTB policies for any trading partner, so we assume that the trade distortions applying in aggregate (estimated from the input–output data) apply in a non-discriminatory fashion to all importers. We could extend this to allow for the discriminatory rates applying to EU member countries following Sweden's recent accession to the EU.

The specification of energy and carbon taxes are central to the model. To capture their structure, particularly with respect to the use of sectoral exemptions, we model them as falling on trade in intermediate inputs. This allows us considerable flexibility to calibrate the model precisely to capture the distortionary effects of existing taxes at the correct margin in terms of our model. Table 5.4 lists the estimates we have generated of the carbon taxes applicable in Sweden in 1995, and Table 5.5 lists the estimates for energy and sulphur taxes. These rates are displayed as follows: each column shows the good whose use as an input in the production of the row good generates the percentage

Table 5.4. Benchmark carbon taxes (%).

Purchasing sector	Input		
	STEN	PETR	GASV
JORD	268	64	61
SKOG		59	61
FSKE		66	61
JARN	268	84	61
A_ME	268	90	61
STEN		67	61
SLAK	67	19	15
MEJE	67	20	15
FRUK	67	21	15
FISK	67	19	15
FETT	67	20	15
KVAR	67	18	15
BAGE	67	18	15
SOCK	67	24	15
CHOK	67	22	15
DIVX	67	20	15
FODE	67	21	15
DRYC	67	20	15
TOBA	67	20	15
GARN		22	15
TEXT		17	15
TRIK		18	15
OVRT		19	15
BEKL		15	15
LADE		15	15
S2GV	67	18	15
TRAH	67	18	15
A_TR	67	23	15
OVR_	67	16	15
TRAM	67	17	15
PAPP	67	24	15
PPPP		24	15
TRAF	67	25	15
PFRP	67	21	15
OVRX	67	19	15
GRAF	67	13	15
KEMI	67	20	15
GODS	67	19	15
BASP	67	23	15
PLAS	67	21	15
FARG	67	15	15
LAKE	67	22	15
TVAT	67	15	15
OVRK	67	20	15
PETR		25	15
SMOR	67	21	15

Continued.

Table 5.4. (*Continued.*)

Purchasing sector	Input		
	STEN	PETR	GASV
GUMM	67	18	15
PLSV	67	17	15
PORS		18	15
GLAS		20	15
TEGE		19	15
CEME		20	15
OVRM		19	15
JRN_	67	20	15
FERR	67	20	15
JNGJ	67	21	15
META	67	20	15
METV	67	20	15
I_JA	67	20	15
METR	67	17	15
MSKN	67	16	15
ELMO	67	14	15
TELE	67	17	15
HUSH	67	17	15
OVRE	67	16	15
VARV	67	18	15
RALS	67	18	15
BILA	67	16	15
CYKL	67	18	15
FLYG	67	14	15
OVRR	67	18	15
INST	67	12	15
A_TI		16	15
EL_O		87	61
GASV	268	87	61
VATT		61	
BYGG		58	61
VARU		55	61
HOTE		55	61
SAMF		66	61
POST		55	61
BANK		55	61
EGNA		76	61
FAST		76	61
UPPD		55	61
REPA		55	61
OVRP	268	55	61

tax liability indicated.[10] Thus, for example, production in sector JORD uses intermediate inputs from sector PETR and effectively incurs an *ad valorem* carbon tax of 64% on those inputs. Similarly, sector JORD uses inputs from

Table 5.5. Benchmark energy and sulphur taxes (%).

Purchasing sector	Input		
	STEN	PETR	GASV
Energy taxes			
JORD	77	108	16
SKOG		117	16
FSKE		111	16
JARN	77		16
A_ME	77		16
STEN	77		16
EL_O	77	68	16
GASV	77	68	16
VATT		16	
BYGG		112	16
VARU		110	16
HOTE		110	16
SAMF		109	16
POST		110	16
BANK		110	16
EGNA		70	16
FAST		70	16
UPPD		110	16
REPA		110	16
OVRP	77	110	16
Sulphur taxes			
JORD	56	0.5	
SKOG		0.2	
JARN	56	5	
A_ME	56	7	
STEN	56	0.3	
EL_O		7	
GASV	56	7	
BYGG		0.1	
SAMF		3	
EGNA		2	
FAST		2	
OVRP	56		

sector GASV and pays instead an effective carbon tax of 61%. These estimates take into account the partial exemptions for manufacturing sectors applicable for carbon taxes in 1995. The energy and sulphur taxes in Table 5.5 should be read the same way.

Information on value added taxes, social security taxes on labour, capital taxes, import tariffs, production taxes (other than energy or pollution taxes), and production subsidies are assembled from various sources described in

Harrison and Kriström (1997, Appendix A). The rates assumed for the value added taxes and factor taxes reflect statutory rates applicable in 1995, and the other rates reflect actual collections as documented in the input–output table for 1992. Although these pre-existing distortions are all incorporated at a detailed sectoral level, in many cases the sectoral variations are small. This feature of the model could be improved with additional work on the background data, and would likely result in more substantial second-best effects from the carbon tax scenarios considered later.

Estimates of elasticities of substitution must be assumed for primary factor substitution, value added and intermediate input substitution, import demand, detailed import components, import source, and domestic demand; elasticities of transformation must also be assumed for the allocation of domestic supply into domestic and exported markets, the allocation of exports into detailed export components, and the allocation of exports to destination. Despite our literature search, there are many elasticities about which there is considerable uncertainty. Our solution for that problem is to undertake a systematic sensitivity analysis as described in Harrison and Kriström (1997) with respect to key elasticities. Harrison and Vinod (1992) and Harrison et al. (1993a, 1993b) demonstrate the role of systematic sensitivity analysis of models such as these with respect to plausible ranges of uncertainty about key elasticities.

The trade elasticities assumed in the model are particularly important. Higher trade elasticities tend to result in greater substitution away from energy-intensive sectors in Swedish production, as untaxed foreign production is substituted for taxed domestic production. We therefore use trade elasticities that reflect the best econometric estimates currently available (Reinert and Roland-Holst, 1992; Reinert and Shiells, 1991). Although they are low in relation to elasticity estimates used in some modelling exercises (e.g., Harrison et al. 1995, 1996, 1997), it is important to stress that they are based on explicit econometric estimates, and used in a model that rules out any 'terms of trade effects' by assumption.[11]

Estimates of carbon emissions in each sector were derived on the basis of information on physical usage of primary energy inputs. These data can then be used to infer the amount of CO_2 generated by each sector, since emissions are a reliable multiple of the physical amount of primary energy used. These estimates are listed in Table 5.6 for each sector, and reveal a familiar structure of the carbon economy. The biggest emissions in aggregate terms come from SAMF (transport), EL_O (electricity generation), and the iron and steel complex (sectors JRN_, FERR, JNGJ, META, METV, and I_JA). Between them these sectors account for 71% of total domestic emissions.

Another measure of the dirtiness of a sector can be obtained by the level of carbon emissions for each million SEK of output it produces. By this measure the iron and steel complex comes off much worse than the transport and electricity sectors, generally by an order of magnitude.

Table 5.6. Carbon emissions in the Swedish model.

Sectors	Aggregate emissions (1000 tons)	Percent of domestic emissions	Rank of percent emissions	Cumulative percent	Emissions per billion SEK output	Rank of per unit emissions
JORD	1388	3	10	77	29	21
SKOG	424	1	21	88	16	25
FSKE	192		34	95	96	11
JARN	266	1	27	92	73	13
A_ME	214		29	93	70	14
STEN	70		47	97	3	62
SLAK	113		38	96	3	64
MEJE	118		36	95	5	48
FRUK	57		52	98	5	47
FISK	39		61	99	5	44
FETT	38		62	99	6	40
KVAR	35		65	99	7	35
BAGE	98		41	96	6	42
SOCK	116		37	96	48	17
CHOK	57		53	98	6	39
DIVX	60		48	98	6	41
FODE	47		56	98	7	36
DRYC	98		42	97	3	59
TOBA	33		67	99	2	69
GARN	59		51	98	8	32
TEXT	17		80	100	3	66
TRIK	18		78	100	1	81
OVRT	19		77	100	3	65
BEKL	7		84	100		86
LADE	3		86	100		85
S2GV	98		40	96	4	51
TRAH	45		58	99	5	49
A_TR	27		70	100	6	38
OVR_	6		85	100	2	77
TRAM	28		69	99	1	79
PAPP	434	1	20	87	32	20
PPPP	398	1	22	89	9	30
TRAF	368	1	23	90	646	4
PFRP	95		43	97	15	26
OVRX	79		45	97	6	43
GRAF	105		39	96	2	70
KEMI	74		46	97	3	55
GODS	16		81	100	5	46
BASP	39		60	99	3	63
PLAS	53		54	98	5	45
FARG	21		73	100	3	57
LAKE	60		49	98	3	58
TVAT	17		79	100	2	71
OVRK	48		55	98	3	56
PETR	81		44	97	1	80

Continued.

Table 5.6. (Continued.)

Sectors	Aggregate emissions (1000 tons)	Percent of domestic emissions	Rank of percent emissions	Cumulative percent	Emissions per billion SEK output	Rank of per unit emissions
SMOR	59		50	98	16	24
GUMM	23		72	100	2	72
PLSV	46		57	99	3	61
PORS	179		35	95	65	15
GLAS	326	1	24	90	41	19
TEGE	213		30	93	147	8
CEME	225		28	92	166	7
OVRM	326	1	25	91	24	22
JRN_	2404	5	3	49	64	16
FERR	2404	5	3	49	1565	1
JNGJ	2348	5	5	58	1220	3
META	2170	4	6	63	264	5
METV	2160	4	7	67	217	6
I_JA	2155	4	8	71	1289	2
METR	208		33	94	3	68
MSKN	211		31	94	2	76
ELMO	20		74	100	2	75
TELE	33		68	99	1	83
HUSH	12		82	100	2	73
OVRE	37		63	99	1	78
VARV	37		64	99	5	50
RALS	33		66	99	8	31
BILA	209		32	94	2	74
CYKL	19		76	100	7	34
FLYG	44		59	99	3	67
OVRR	20		75	100	22	23
INST	24		71	100	1	82
A_TI	8		83	100	1	84
EL_O	9622	19	2	44	136	9
GASV	266	1	26	91	134	10
VATT		87	100		87	
BYGG	1474	3	9	74	8	33
VARU	538	1	12	79	47	18
HOTE	538	1	12	79	12	28
SAMF	12352	25	1	25	80	12
POST	538	1	12	79	12	29
BANK	538	1	12	79	7	37
EGNA	459	1	18	85	4	54
FAST	459	1	18	85	4	52
UPPD	538	1	12	79	3	60
REPA	538	1	12	79	14	27
OVRP	538	1	11	78	4	53
Total	50029	100				

Table 5.7. Simulation scenarios.

BENCH	Maintain all policies at their initial level and replicate the benchmark economy.
C100	Increase the existing structure of carbon taxes in Sweden by 100% above their benchmark rates, maintaining the existing exemptions from carbon taxes. Reduce labour taxes to maintain constant government revenue.
DIES	Increase the diesel tax so as to match the petrol tax in terms of carbon emissions. Reduce labour taxes to maintain constant government revenue.
PETROL	Double the petrol tax. Reduce labour taxes to maintain constant government revenue.

Comparing the estimates of carbon taxes and the estimates of carbon emissions, the absence of taxes on the iron and steel complex is immediate. The formal reason for this is that these sectors are exempt. The stated rationale underlying this exemption is that they are particularly vulnerable to foreign competition and would be unable to pass on any taxes on one of their inputs unless their competitors also bore comparable taxes.

Another feature of this comparison of sectoral taxes and sectoral emissions is that, of the two biggest aggregate emitters (SAMF and EL_O), only EL_O pays any tax on inputs of coal (output from sector STEN). Moreover, this tax is levied as an energy tax, and not as a carbon tax. Thus one could imagine the incentive within that sector to move away from coal-fired generators as the result of scalar increases in energy taxes. This margin of choice is incorporated in the model, to the extent that sector EL_O can substitute away from intermediate inputs of STEN and towards PETR (or, to a lesser extent, GASV and SMOR).[12] The current version of the model adopts a CES production technology with respect to intermediate inputs, and assumes an elasticity of substitution of 0.25. It would obviously be useful to consider richer specifications of the energy technology in sector EL_O in future work.

The SOE model is generated with the GAMS/MPSGE software developed by Brooke et al. (1992) and Rutherford (1992, 1995). It is then solved using the MILES algorithm developed by Rutherford (1993) or the PATH algorithm developed by Dirkse and Ferris (1995). Harrison and Kriström (1997, Appendix B) document the computer software in some detail. Each scenario typically solves in less than a minute on a Pentium-based personal computer running at 90 mHz with at least 16 mb RAM.

5.3. Effects of policies

5.3.1. Baseline policies and simulation scenarios

Table 5.7 lists the simulations we report here. The core simulation, which we then interpret with the other simulations, is called C100 and involves a 100%

increase in existing carbon taxes in Sweden. As a default we lower labour taxes so as to ensure equal government revenue after the carbon tax policy. Thus C100 incorporates the existing structure of carbon taxes, in particular the current exemptions.

In order to describe the DIESel and PETROL scenarios in some detail, it is useful to review the current energy taxes on these fuels. Assuming a net price of 1468 SEK/m^3 for diesel and a net price of 2175 SEK/m^3 for petrol, the energy tax imposes a percentage increase of 109% and 148% on diesel and fuel. By contrast the carbon tax imposes a percentage increase of only 67% and 36%, respectively.[13]

The primary purpose of the energy tax is to raise revenue. The carbon tax, on the other hand, reflects an underlying tax of 0.34 SEK/kg CO_2 and is designed to meet an explicit environmental goal. Because diesel and petrol are roughly comparable in terms of kWh/m^3, it is apparent that diesel has a much lower energy tax. When converted to SEK/kWh we obtain an energy tax of 0.17 on diesel and 0.4 on petrol. The corresponding carbon tax in these terms if 0.11 and 0.1, which is much more uniform.[14]

The purpose of the DIESel scenario is to study the impact of raising the energy tax for diesel such that diesel and petrol has the same energy tax (in terms of kWh). This means that the tax of diesel changes from 109% to 248% in the simulations. This represent the intuitively appealing idea of making the tax system more symmetrical.

The PETROL scenario simply involves doubling the price of petrol. This policy is consistent with one of the key proposals to be floated in Sweden by the Traffic and Climate Committee (Statens Offentliga Utredningar, 1995), as discussed earlier.

5.3.2. *Effects of expanding the carbon tax*

Welfare impacts

The detailed welfare impacts of the C100 scenario are presented in Table 5.8. The first column lists the acronym of the household, defined in Table 5.2. The second and third columns report the percentage share of each household type in the total population of households or individuals.[15] We can use households or individuals as the bases of alternate social welfare function. Using individuals has the effect, relative to using households, of giving the single person household groups a lower weight in social welfare, and enhances the weight of those households with more children.

The fourth column reports the value of the utility index for each household, normalized without loss of generality to 100 in the benchmark. Thus a value of 99.7 in this column indicates that the household type has experienced a decrease in the utility index of 0.3%. A more meaningful evaluation is provided in the final two columns, which list the equivalent variation (EV) in income

Table 5.8. Welfare impact of doubling the carbon tax (scenario C100).

Household	Percentage share of		Utility index	EV in SEK per	
	Households	Individuals		Individual	Household
S_NC_1	9.2	4.2	99.7	−415.0	−415.0
S_NC_2	9.1	4.2	99.8	−283.0	−283.0
S_NC_3	9.1	4.2	99.8	−409.0	−409.0
S_NC_4	9.2	4.2	99.9	−345.0	−345.0
S_C_1	1.8	1.9	99.8	−226.0	−521.0
S_C_2	1.8	2.2	99.7	−431.0	−1164.0
M_NC_1	7.3	6.7	99.5	−617.0	−1234.0
M_NC_2	7.4	6.8	99.7	−464.0	−928.0
M_NC_3	7.3	6.7	99.6	−653.0	−1307.0
M_NC_4	7.3	6.7	99.7	−693.0	−1387.0
M_1C_1	1.9	2.6	99.6	−386.0	−1157.0
M_1C_2	1.9	2.5	99.6	−448.0	−1343.0
M_1C_3	1.9	2.6	99.6	−537.0	−1611.0
M_1C_4	1.9	2.6	99.6	−762.0	−2287.0
M_2C_1	2.4	4.4	99.5	−416.0	−1666.0
M_2C_2	2.4	4.4	99.5	−481.0	−1924.0
M_2C_3	2.4	4.4	99.4	−602.0	−2407.0
M_2C_4	2.4	4.4	99.4	−758.0	−3033.0
M_3C_1	1.1	2.5	99.6	−299.0	−1557.0
M_3C_2	1.1	2.5	99.6	−363.0	−1887.0
M_3C_3	1.1	2.5	99.5	−461.0	−2397.0
M_3C_4	1.1	2.6	99.5	−537.0	−2900.0
O_NC_1	1.4	1.4	99.7	−436.0	−959.0
O_NC_2	1.4	1.7	99.7	−418.0	−1128.0
O_NC_3	1.4	1.8	99.7	−542.0	−1571.0
O_NC_4	1.4	2.1	99.7	−562.0	−1911.0
O_C_1	0.9	1.6	99.7	−348.0	−1323.0
O_C_2	0.9	1.8	99.7	−320.0	−1375.0
O_C_3	0.9	1.8	99.7	−456.0	−1963.0
O_C_4	0.9	1.9	99.6	−569.0	−2562.0

needed to make the individual or household as well off as they are in the new counter-factual equilibrium (evaluated at benchmark prices).

The EV is positive for welfare gains from the counter-factual policy scenario, and negative for losses. We report it in terms of SEK over a one year period for each individual in the household group or for each household in the household group. Thus these values can be interpreted as the minimum amount of money that each individual or household in each household group would need to have received, if the policy or scenario had not occurred, for them to just as well off as if it had occurred. It is important to note that this welfare evaluation takes no account of the direct benefits to the household of the resulting reduction in aggregate emissions of either pollutant. Thus we can view these estimates as indicators of the minimum benefits which each consumer would have to perceive from the reduction in pollution in order for that consumer to regard the policy as a good one from an individual perspective.

In the C100 scenario we can therefore see that all household groups lose from a doubling of the existing carbon tax. For the single-adult household the cost is relatively modest, and well below the cognitive threshold value of 500 SEK. The costs become more substantial for all other households, especially those with children. Married households with no children experience slightly higher costs than single households with no children. In general richer households within any group tend to bear higher costs, reflecting the greater carbon-intensity of their expenditure patterns and their higher initial incomes.[16]

There is an intriguing effect of having extra children on the costs of the carbon tax increase for households. Having one or two children tends to raise the cost to a married household. But having three or more children actually reduces the household cost. The puzzle is resolved by examining how expenditure patterns change with extra children, not to mention some introspection.[17] Having children implies that households must use consumption technologies that have a significant fixed cost component: the purchase of durables such as prams and toys. These tend to be more carbon-intensive than the variable cost component of having children (i.e. toys actually have more embodied carbon-content than diapers), and it is the variable cost component that plays more of a role for the second child since the fixed cost expenditures do not have to be as large. The effect from having more than one child appears to be due to an increase in the share of household expenditures being allocated to transport. Presumably this reflects the need to take more family holidays, or the effects of relocation decisions as households tend to move out of dense (and carbon-efficient) urban transport networks into suburban transportation networks.[18]

The costs of the carbon tax increase is greatest for households that are married with two children, and for richer households. The other households group also tends to bear a relatively high burden; this group consists mainly of children above the age of 17 living at home with their parents.[19] These households experience losses that are generally greater than 1000 SEK/year, and in several cases are more than 2500 SEK/year.

To repeat an important point, the fact that all households experience a loss does not mean that they would not benefit overall from the carbon tax increase. The reason is that we have neglected the direct benefit they would reap from the reduction in aggregate carbon emissions that would (presumably) result from the policy. In fact our model estimates that there would be a reduction of CO_2 of 52 ktons, as discussed later.[20] Although this is a modest reduction in percentage terms, it is possible that household M_2C_4 would value it at more than the 3033 SEK/year that would be the cost to that household to bring about the reduction. In the absence of any formal attempt to estimate the direct benefits to Swedish households from carbon reductions of various magnitudes, such judgments will have to be made politically. We provide some guidance on this matter later, but do not pretend that we know what these gross benefits are.

It should also be added that different households might have very different perceptions of the direct benefits of carbon reductions. Hence it could be the case that household M_2C_4 does get a benefit that exceed the price it pays of 3033 SEK, but that household S_NC_2 does not benefit by more than the modest price of 283 SEK which it must pay. The gross benefits of any given commodity, whether it be 'stor stark öl' or '52 less ktons of carbon on the planet', can vary from household to household and individual to individual. Indeed, it is plausible that having more children would make one more concerned about the quality of the environment in the future, and increase one's willingness to pay for carbon reductions. On the other hand, having children may also increase your discount rate, such that the enhanced benefits of carbon reduction in the future are insufficient to offset the increased price to be paid now.

This is not to say that our estimates of welfare costs are worthless, but simply to identify the many factors which must be considered before they can be properly used to guide decision-making. Implicit or explicit estimates of discount rates and gross benefits from carbon reductions must be made before an overall assessment of the C100 policy is possible. We stress these considerations since we will generally proceed to ignore them when describing the results.

There are several ways in which to aggregate these detailed welfare impacts. The first is to just add up the EV values for all households, ignoring the distributional impact. In effect this represents the evaluation one gets from a simple utilitarian social welfare function (SWF). This type of SWF ignores who gains and loses, and only focuses on whether the aggregate pie has increased or not. In the present case it has clearly decreased, and the aggregate loss in income is 4 billion SEK per year. This aggregate is obtained by adding up the EV values in either of the last two columns of Table 5.8, multiplying each by the number of individuals or households in the household type as appropriate. It openly ignores the distributional burden of the welfare impacts.

Another way in which the overall impact of the C100 policy could be viewed is that it is the aggregate price tag for the Swedish economy of a reduction in emissions of 52 ktons CO_2. A social counterpart to the more complete cost–benefit calculation described above for each individual household could now be undertaken. Such a calculation would require an estimate of the aggregate social benefits to Sweden of this reduction in physical emissions, perhaps by some official body such as the Green Tax Commission. This calculation would again entail the implicit or explicit use of a discount rate, in this case the social discount rate.

Emissions impacts

How did we arrive at the estimate that a reduction of 52 ktons of CO_2 would result from the C100 policy? The sectoral impact shown in Table 5.9 shows how these estimates were arrived at. Consider the last three columns, which show the aggregate change in physical emissions of CO_2 attributable to each sector.

The first of the three columns, marked CO2_D, shows the change due to changes in domestic production in that sector brought about by the C100 scenario. Thus we see that a reduction in domestic production of the JARN sector, indicated by a 1% reduction in the value of domestic value added in column VA%, led to a reduction in physical emissions from that sector of 4 ktons.

The fact that some sectors expand when there is an increase in carbon taxes is exactly what one would expect from a general economic equilibrium. The doubling of the carbon tax changes relative prices against the most carbon-intensive activities. The cheapest way for some industries to contract their use of the (intermediate) inputs of these carbon-intensive sectors may be to substitute towards the use of the products of other sectors that, while less carbon intensive than the ones they displace, might still be more carbon intensive than average for the economy as a whole. Why don't they substitute towards the products that are least carbon-intensive? Simple: their existing technology may not call for them to be used at all. So, even if they have the best relative price ratio because of the carbon tax hike, the value of their marginal product (as inputs) is still virtually zero.

For example, the DRYC sector is a wonderful sector, justifiably patronized by many Swedes. It also has a relatively low (direct) carbon intensity of only 3 ktons carbon/billion SEK of output. But when some sector such as JORD is contemplating increased prices for transportation, and hence increased prices for the inputs of manufactured transport equipment in sectors RALS, BILA and FLYG in our model, it cannot 'turn to DRYC' despite the temptation. It must re-allocate amongst these three transportation input sectors, and in fact such decisions tend to go against RALS and in favour of the other two. The common sense reason that DRYC does not get the nod is that it has nothing technologically to do with reality-based transportation. The formal counterpart of this sobering intuition in our model is that the JORD sector has virtually no (direct) inputs of DRYC in the benchmark year of our input–output table, but it has substantial inputs of all three of the transportation inputs. Hence, by Marshall's second law of derived demand[21] the elasticity of demand for the alternative transport inputs will be relatively large and we can expect to see some net substitution effects there. Conversely, the elasticity of demand for DRYC will be relatively low, so we will not see any changes in the derived demand for it, despite it having a relatively favourable price ratio compared to transport inputs.

Turning now to the next to last column in Table 5.9, CO2_F, we see the

Table 5.9. Sectoral impact of doubling the carbon tax (scenario C100).

Sector	IPRICE%	VA%	IMP%	EXP%	CO2_D	CO2_F	CO2_W
JORD				−1	−1	1	
SKOG	−1		−1				−1
FSKE				−1	−1		−1
JARN		−1		−2	−4		−4
A_ME				−1			
STEN		−6	−7	−7	−5	−8	−13
SLAK	−1						
MEJE				−1			
FRUK	−1						
FISK	−1						
FETT	−1						
KVAR	−1						
BAGE	−1						
SOCK	1			−1			
CHOK	−1						
DIVX	−1						
DRYC	−1						
TOBA	−1	1					
GARN	−1						
TEXT	−1	1	−1				
TRIK	−1	1		1			
OVRT	−1	1					
BEKL	−1	1					
LADE	−1	1					
S2GV	−1		−1				
TRAH				−1			
A_TR	−1						
OVR_	−1						
TRAM	−1						
PAPP			−1	−1	−2		−2
PPPP				−1	−2		−1
TRAF	−1				1		1
OVRX	−1	1					
GRAF	−1						
KEMI				−2			
GODS				−1			
BASP	−1						
PLAS	−1	1					
LAKE	−1	1		1	1		1
TVAT	−1						
OVRK	−1		−1				
PETR	18	−9	−2	−23	−11		−11
SMOR			−1	−1			
GUMM	−1						
PLSV	−1	1					
PORS	−1				1		1
GLAS							1

Continued.

Table 5.9. (*Continued.*)

Sector	IPRICE%	VA%	IMP%	EXP%	CO2_D	CO2_F	CO2_W
TEGE				−1	−1		
CEME			−1	−1	−1		−1
OVRM				−1			
JRN_				−1	−6	1	−4
FERR	−1				3	−1	2
JNGJ	−1	1			11	2	13
META					4	2	6
METV	−1				7	4	11
I_JA	−1	1			11	2	13
METR	−1						1
MSKN	−1	1			1		1
ELMO	−1	1					
TELE	−1	1		1			
HUSH	−1	1		1			
OVRE	−1	1					
VARV	−1						
RALS	−1		−1				
BILA	−1	1		1	2		2
CYKL	−1	1		1			
FLYG	−1		−1				
OVRR	−1						
INST	−1	1					
A_TI	−1						
EL_O				−1	−5		−5
GASV	16	−1			−13		−13
BYGG	−1				1		1
VARU			−1	−1	−2		−2
HOTE	−1				1		1
SAMF				−1	−57		−57
POST	−1				1		1
EGNA	−1	1			3		3
FAST	−1				1		1
UPPD	−1				1		1
REPA	−1				2		2
OVRP	−1	1			2		3
Total					−52	6	−47

effect of the Swedish policy on foreign emissions of CO_2.[22] Virtually any domestic policy is going to have some impact on the structure of Swedish imports, as changes in the relative prices of domestic goods cause Swedes to substitute in favour of or against foreign goods. In the present case there will be substitution away from those goods whose input price, shown as percentage change in Table 5.9 in column IPRICE%, has increased. The clearest instances are as expected, PETR and GASV. In each case there is a large increase in domestic prices brought about by the doubling of the carbon tax: after all of the general equilibrium effects have worked themselves out, the final domestic price increase is about 18% or 16%. This results in a fall in domestic production, and a switch towards imports, shown as a percentage change in Table 5.9 in column IMP%. There is also a reduction in exports, shown in column EXP% in Table 5.9, for the same reason: Swedish exports in these carbon-intensive goods are simply unable to compete with foreign goods at (unchanged) world prices.

Hence we have an increase in the value of foreign imports of SMOR, and indeed in the physical quantity of imports. If we were to assume that foreign producers are as carbon-efficient as Swedish producers in the same industry, then there would be an increase in carbon emissions overseas due to the increased foreign production needed to meet Sweden's increased import demand. In fact we assume that foreigners are not as carbon-efficient as Sweden, which is generally a plausible assumption apart from extremely nuclear-intensive countries. The exact assumptions as to how much 'dirtier' foreign production is[23] are not so important as the general logic that accounts for the foreign change in emissions. That logic is important since it is global emissions that matter for the final environmental good, reduced risk of global warming. Hence it is incumbent on Sweden to take into account the leakage effects of just reducing on-shore carbon-intensive activities and substituting off-shore production of those products.

We acknowledge that we do not undertake a full multi-regional evaluation of this leakage issue, and there are obvious limitations to calculations of this kind. It is possible that changes in Sweden's exports will change production patterns overseas in ways that could increase or decrease carbon emissions globally. More generally, since we do not model the general equilibrium of foreign economies, we are not accounting for the full effects of changes in Sweden's net trade pattern. Given these qualifications, which are inherent to the use of a single economy model, we believe it is important to acknowledge the potentially offsetting effects of carbon tax reforms when international trade is taken into account. There are, of course, many sectors where the foreign effect works in the same direction as the domestic effect (e.g. STEN), so our incorporation of foreign effects should not be viewed as imparting a presumptive bias into the estimation of global emissions.

The final column in Table 5.9, CO2_W, shows the aggregate world change in emissions of carbon in each sector. The foreign effects tend to be dominated by the domestic change, since imports are generally a much smaller of domestic consumption in most sectors than domestic production.

Price and production impacts

The evaluation of welfare impacts and emissions impacts are, in an important sense, the bottom line of our policy simulations since they provide the ultimate basis for evaluating the policy. Examining them provides an idea of what is happening to the Swedish economy as the result of the C100 policy. However, it may be useful to look more directly at the changes in prices, production and trade to see the underlying causes of these effects.

From the IPRICE% column in Table 5.9 we see that the PETR and GASV sectors face a large price increase. Given the structure of carbon taxes, as shown in Table 5.4, these first order impacts are not surprising.

Why do prices for PETR and GASV, however, only rise by about 17% when the *ad valorem* rates of carbon taxes listed in Table 5.4 appear to be anywhere from 15% up to 90%? The answer is to recall that the higher rates do not apply to all sectors that use PETR and GASV, particularly energy-intensive manufacturing sectors. Thus if we average out the carbon tax rate on PETR and GASV over all sectors, including those that are exempt from it and are not listed in Table 5.4, the average rate would be closer to the observed price changes. In addition, the final price changes shown in Table 5.9 will reflect additional second-order impacts due to resource re-allocations by consumers, producers and foreigners. Nonetheless, we would expect the first-order effects on prices to dominate for a scenario like this one.

Why is there such a small impact on the price of electricity, sector EL_O? Indeed, there is a slight increase in the price of sector EL_O, but it does not round up to 1% and hence is shown as a blank in our reports. Nonetheless, why is there not a larger increase, since EL_O has to be carbon-intensive? The immediate response is that Swedish electricity generation is dominated by nuclear and hydro, which are not carbon-intensive; that sector EL_O includes district heating, which is not carbon-intensive; and that sector EL_O is exempt from carbon taxes on the use of coal.

Essentially the same answer to this question comes from considering in detail the use of intermediate inputs that are hit with the carbon tax, and then seeing what happens to their prices. Since we know that PETR and GASV have substantial price increases, the implication of a small price increase for EL_O is that it must not use very much of these as intermediate inputs. It is instructive in the economics of our model to work this issue through further.

Sector EL_O has five sources of primary energy inputs in our model.[24] Three are those listed in the columns of Table 5.4 as bearing carbon taxes: STEN, PETR and GASV. The fourth is SMOR, which does not bear any carbon taxes. The fifth is EL_O itself, which is where all of the nuclear-generated primary energy comes from in the input–output database. Of these five intermediate inputs, the cost shares in 1992 were: STEN 39%, PETR 26%, SMOR 0%, GASV 15% and EL_O 20%, while it would still seem that the taxes on STEN, PETR and GASV should affect EL_O prices, these percentages are misleading as to the complete cost structure of the EL_O sector. For example, the EL_O sector spent about as much on consulting and lobbying services (uppdragsverksamhet, or UPPD) in 1992 as it did on PETR, and while consultants and lobbyists obviously generate a lot of negative externalities they are not (yet) subject to any pollution tax.

As a share of total intermediate inputs, then, the cost shares in 1992 were much smaller: STEN 11%, PETR 8% and GASV 5%. A simple piece of arithmetic suggests that the weighted carbon tax on EL_O from these three inputs is only 7.13% = (0% × 0.11) + (87% × 0.08) + (61% × 0.05). However, even this calculation overstates the effective tax in our model and the economy, since there are some possibilities for EL_O to substitute away from the more heavily taxed input PETR, and indeed away from all of the taxed inputs, since there are other inputs used in the benchmark technology to produce its output.[25]

5.3.3. *Effects of the petrol tax and diesel tax proposals*

The tax revenues are used to reduce labour taxes, as described earlier, such that there is no net revenue effect on government. There is, nevertheless, a welfare loss of 0.4–3 billion SEK, which is equivalent to about 100 SEK and 750 SEK per household, respectively. Using an exchange rate of US$1 ≈ SEK 6.50, the equivalent variations are roughly US$15 and US$115 per household. The second column of Table 5.13 presents the aggregate welfare loss in percentage terms, where the DIESel scenario is associated with very small percentage losses (less than 0.1%).

The higher diesel tax primarily hits the transportation sector. It is important to note the possibility of tax leakages in this context, given the fact that about 9% of the total distance driven by trucks (above 3.5 ton) in Sweden are foreign. Because the price of diesel is generally higher in Sweden than in neighbouring countries (Norway is an exception), it is likely that the tax leakage effect is small. Foreign vehicles are probably not using Swedish diesel, although we have no data to confirm this intuition.

The differences between diesel prices in neighbouring countries suggest increased incentives for border trade. In order to determine the amount of tax leakage through this channel, it is useful to note that Swedish trucks tend to drive short distances. Indeed, only 6% of all journeys cover distances exceeding 300 km. Consequently, the price increase is going to hit mainly Swedish trucks and while there is some possibility for border trade (mainly the Swedish–Finnish border in the north), the tax leakage is likely to be quite small.

The environmental impacts are small according to the model. Neither scenario involves drastic reductions of the CO_2 emissions. The higher diesel tax leads to a minute increase of foreign emissions through increased imports. These effects are so small as to be negligible. It is important to realize that CO_2 emissions might increase in some sectors, because CO_2 emissions are modelled via fixed coefficients on sectoral output, and expansion of some sectors implies an increase of their emissions. In other words, we do not allow in the present version of the model for the possibility that pollution mitigation expenditures might be employed to reduce the rate at which carbon is emitted in relation to output. This mitigates some of the decreases one expects from sectors that are heavy users of diesel, such as the transportation sectors.

Similarly, there is no detailed modelling of the possible substitution between diesel and petrol. We do allow firms to substitute intermediate inputs whose prices might vary due to changes in the relative price of diesel and petrol, to the extent that this inputs are used in benchmark data, but these possibilities are limited. We also allow consumers to substitute away from final goods that experience a relative price increase due to the relative intensity of diesel and petrol use in their production, so some indirect substitution between diesel and petrol can occur at this level. Again, these possibilities are relatively limited. What would be needed is a detailed model of how each industry chooses which of fuel to use (and how it then chooses the vintage composition of its capital stock to accommodate those fuel choices).

Taken together these features of the model suggest that the actual reductions of CO_2 might be larger than suggested by our simulations.

5.3.4. *A cost–benefit comparison*

Our model is constructed to generate estimates for each household of the price tag or cost of increases in taxes directed at reducing CO_2 emissions. Is it possible to relate these, even roughly, to estimates of gross benefits from carbon tax reductions? Although proper gross benefit estimates do not exist for Sweden, or indeed for any country, there have been some estimates floated in international circles that can be usefully related to our cost estimates.

The source for these gross benefit estimates is the Inter-Governmental Panel on Climate Change (IPCC), specifically Working Group 3.[26] Based on some loose avoided cost calculations, they tentatively offer US$125/ton carbon as an upper bound on gross benefits. We carefully translate that into ktons/CO_2 for comparison with our model, and then into SEK from US$.

The IPCC report does not indicate whether they intend this number to refer to individuals or households, so we apply it to both. The IPCC report also does not say if this estimate is an aggregate over individuals or households, or is meant to be interpreted per individual or per household. Since the underlying avoided cost calculations are aggregative in nature, we assume that this estimate applies as an aggregate. To be conservative, we further assume that it applies to the aggregate population (of individuals or households) in Sweden, and not the planet. We then apportion the benefits proportionally across households, according to that household's share of the aggregate number of individuals or households. This assumption is appropriate given that we have no priors or data to suggest that one household group would value carbon reductions any greater than another.

We further assume that this gross benefit estimate is linear in the kton reduction in CO_2 that our model generates for any particular scenario. In the case of C100, for example, we estimate a 52.2 kton reduction, so we are in effect assuming that each household receives the same gross benefit from the first kton reduction as from the last. Although we might justify such an assumption based on the small scale of this carbon reduction, and hence the approximate linearity of the unknown marginal benefit schedule, our primary concern is to keep the arithmetic simple and transparent. It should not be assumed that marginal benefit would decline, due to diminishing marginal utility arguments, since households may correctly perceive the importance of threshold effects in carbon reductions. In other words, I might be willing to pay nothing for small decreases in carbon emissions, but substantially more if I perceive that the aggregate emission reduction might make a difference to the risk of global warming. Our cost estimates do, however, take into account the non-linearity of the underlying preferences and technologies for larger and larger reductions in emissions.

The resulting estimates for each household in scenario C100 are presented in Table 5.10. Comparable estimates for the PETROL scenario are shown in Table 5.11. In each case the last row shows the average benefit and cost over all households, and each row shows the arithmetic for each household. We use an estimate of the gross benefit which is actually double the upper bound of the IPCC estimate, so as to avoid any risk of understating those benefits.

The conclusion is clear. The benefits[27] of doubling the carbon tax or the petrol tax in Sweden are a tiny fraction of the price tag which Swedes must pay in the form of higher prices and reduced incomes. The results for the DIESel simulation are comparable, with average estimates of the cost to individuals of 53 SEK (115 SEK), and benefits that do not amount to 0.5 of one SEK. Although we do not put much credence in any of these gross benefit numbers, they do serve to highlight the basis of our conclusion that carbon, diesel or petrol tax increases are not currently justifiable in Sweden. They also serve to focus the debate on the net benefits of further carbon, diesel or petrol taxes onto the question of estimating gross benefits for Swedes. If these numbers are correct, then advocates of these tax increases are telling the average Swede that he or she must pay a lot more for some environmental good than that Swede appears to derive as a benefit. This might be because the advocate derives significant enough benefits and would be willing to pay the price tag, but that does not justify foisting the price on others.

5.4. Conclusions

Our most important conclusion is that unilateral increases in carbon taxes, diesel taxes and petrol taxes do not appear to generate emissions reductions that are sufficient to justify the cost they impose on Swedes. While our model might underestimate the reductions in emissions, it is well known that the relevant price elasticities are small. In particular, the short-term price elasticity of petrol is small. In addition, the average cost share of fossil fuels is small,

Table 5.10. Costs and benefits to Swedes in SEK of doubling the carbon tax (scenario C100).

Household	Average individual			Average household		
	Benefit	Cost	Percent	Benefit	Cost	Percent
S_NC_1	3	415	1	6	415	2
S_NC_2	3	283	1	6	283	2
S_NC_3	3	409	1	6	409	2
S_NC_4	3	345	1	6	345	2
S_C_1	3	226	1	6	521	1
S_C_2	3	431	1	6	1164	1
M_NC_1	3	617		6	1234	1
M_NC_2	3	464	1	6	928	1
M_NC_3	3	653		6	1307	
M_NC_4	3	693		6	1387	
M_1C_1	3	386	1	6	1157	1
M_1C_2	3	448	1	6	1343	
M_1C_3	3	537	1	6	1611	
M_1C_4	3	762		6	2287	
M_2C_1	3	416	1	6	1666	
M_2C_2	3	481	1	6	1924	
M_2C_3	3	602		6	2407	
M_2C_4	3	758		6	3033	
M_3C_1	3	299	1	6	1557	
M_3C_2	3	363	1	6	1887	
M_3C_3	3	461	1	6	2397	
M_3C_4	3	537	1	6	2900	
O_NC_1	3	436	1	6	959	1
O_NC_2	3	418	1	6	1128	1
O_NC_3	3	542	1	6	1571	
O_NC_4	3	562	1	6	1911	
O_C_1	3	348	1	6	1323	
O_C_2	3	320	1	6	1375	
O_C_3	3	456	1	6	1963	
O_C_4	3	569	1	6	2562	
AVE	3	500	1	6	1090	1

which intuitively suggests that the demand reductions will be insignificant in production sectors. Coupling these facts with the result that carbon emissions can actually increase in some sectors due to general equilibrium repercussions, we find support for the model's prediction that the environmental benefits are unlikely to be significant since the emissions reductions are tiny.

We openly admit that we must rely on some heroic assumptions to undertake such a complete cost–benefit calculation, particularly with regards to the gross benefits of emission reductions. However, advocates of these tax increases must

Table 5.11. Costs and benefits to Swedes in SEK of doubled petrol tax (scenario PETROL).

Household	Average individual			Average household		
	Benefit	Cost	Percent	Benefit	Cost	Percent
S_NC_1	2	310	1	5	310	2
S_NC_2	2	233	1	5	233	2
S_NC_3	2	306	1	5	306	2
S_NC_4	2	278	1	5	278	2
S_C_1	2	172	1	5	395	1
S_C_2	2	303	1	5	817	1
M_NC_1	2	409	1	5	819	1
M_NC_2	2	324	1	5	648	1
M_NC_3	2	444	1	5	889	1
M_NC_4	2	471		5	943	1
M_1C_1	2	262	1	5	787	1
M_1C_2	2	296	1	5	888	1
M_1C_3	2	359	1	5	1078	
M_1C_4	2	482		5	1447	
M_2C_1	2	270	1	5	1079	
M_2C_2	2	311	1	5	1243	
M_2C_3	2	381	1	5	1522	
M_2C_4	2	469		5	1874	
M_3C_1	2	201	1	5	1048	
M_3C_2	2	237	1	5	1235	
M_3C_3	2	293	1	5	1524	
M_3C_4	2	337	1	5	1822	
O_NC_1	2	306	1	5	672	1
O_NC_2	2	300	1	5	809	1
O_NC_3	2	374	1	5	1084	
O_NC_4	2	391	1	5	1331	
O_C_1	2	247	1	5	938	1
O_C_2	2	231	1	5	993	1
O_C_3	2	312	1	5	1340	
O_C_4	2	381	1	5	1713	
AVE	2	340	1	5	741	1

also be implicitly making comparably heroic calculations. Our role as modellers is to bring these unstated assumptions into the open, so that they can be rationally debated and evaluated. These results may not be what everyone likes to hear. Since we are not naïve to the political pressures surrounding this issue in Sweden, nor so cynical as to dismiss them as being unworthy of debate, it is incumbent on us to attempt to direct debate on our model and its results into productive areas.

The model is incomplete in terms of a number of important parameters.

Table 5.12. Detailed carbon tax revenue effects of doubling the carbon tax (scenario C100).

Sector	Benchmark revenues (b.SEK)				Change in scenario C100 (b.SEK)			
	STEN	PETR	GASV	TOTAL	STEN	PETR	GASV	TOTAL
JORD	0.496	0.704		1.200	0.495	0.646		1.141
SKOG		0.222		0.222		0.203		0.203
FSKE		0.095		0.095		0.086		0.086
JARN	0.105	0.061		0.165	0.101	0.054		0.155
A_ME	0.013	0.045		0.058	0.014	0.041		0.055
STEN		0.058		0.058		0.047		0.047
SLAK		0.020	0.003	0.022		0.018	0.002	0.020
MEJE		0.019	0.005	0.023		0.017	0.004	0.021
FRUK		0.007	0.009	0.017		0.007	0.009	0.016
FISK		0.002		0.002		0.002		0.002
FETT		0.007	0.003	0.010		0.007	0.002	0.009
KVAR		0.002		0.002		0.001		0.001
BAGE		0.016	0.003	0.020		0.015	0.003	0.019
SOCK	0.009	0.012	0.014	0.035	0.008	0.011	0.013	0.032
CHOK		0.005	0.001	0.005		0.004		0.004
DIVX		0.007	0.001	0.008		0.007		0.007
FODE		0.005	0.001	0.006		0.004	0.002	0.006
DRYC		0.013	0.006	0.019		0.013	0.005	0.018
TOBA				0.001		0.001	0.001	0.002
GARN		0.013	0.001	0.014		0.012	0.002	0.013
TEXT		0.001		0.001		0.001		0.001
TRIK		0.001		0.002		0.002		0.002
OVRT		0.004		0.004		0.003		0.004
BEKL		0.002		0.002		0.002	0.001	0.003
LADE		0.001		0.001		0.001		0.001
S2GV		0.022		0.022		0.021		0.021
TRAH		0.009		0.009		0.008		0.008
A_TR		0.006		0.006		0.006		0.006
OVR_		0.002		0.002		0.002		0.002
TRAM		0.007		0.007		0.006		0.006
PAPP	0.028	0.046		0.074	0.028	0.042		0.070
PPPP		0.107	0.010	0.116		0.097	0.008	0.105
TRAF		0.001		0.001				
PFRP		0.007	0.002	0.009		0.006	0.002	0.008
OVRX	0.001	0.004		0.005		0.004		0.005
GRAF		0.016		0.017		0.015		0.016
KEMI	0.204	0.249	0.001	0.454	0.201	0.226		0.428
GODS	0.045	0.008	0.001	0.054	0.045	0.007	0.001	0.053
BASP	0.003	0.010	0.002	0.016	0.004	0.010	0.002	0.015
PLAS	0.007	0.007	0.002	0.015	0.006	0.006	0.001	0.014
FARG	0.003	0.009	0.001	0.012	0.004	0.007		0.011
LAKE		0.009		0.009		0.008		0.008
TVAT		0.002		0.002		0.001		0.002
OVRK	0.021	0.009	0.003	0.033	0.022	0.008	0.003	0.033
PETR		0.188		0.188		0.123		0.123

Continued.

Table 5.12. (*Continued.*)

Sector	Benchmark revenues (b.SEK)				Change in scenario C100 (b.SEK)			
	STEN	PETR	GASV	TOTAL	STEN	PETR	GASV	TOTAL
SMOR	0.243	0.024	0.002	0.269	0.240	0.023	0.002	0.265
GUMM	0.002	0.004		0.006	0.002	0.004		0.006
PLSV		0.009		0.009		0.008		0.008
PORS		0.004		0.004		0.003		0.003
GLAS		0.024		0.024		0.021		0.021
TEGE		0.004	0.003	0.008		0.005	0.003	0.008
CEME		0.007		0.007		0.007		0.007
OVRM		0.043	0.002	0.045		0.039	0.002	0.041
JRN_	0.145	0.125	0.006	0.275	0.144	0.114	0.005	0.263
FERR	0.003			0.003	0.002	0.001		0.003
JNGJ	0.018	0.003		0.021	0.018	0.003		0.021
META	0.028	0.010	0.002	0.039	0.028	0.008	0.001	0.038
METV		0.007		0.007		0.007		0.007
I_JA	0.013	0.002		0.015	0.013	0.003		0.015
METR	0.002	0.053	0.002	0.057	0.002	0.048	0.002	0.053
MSKN	0.006	0.048	0.002	0.057	0.006	0.045	0.002	0.054
ELMO		0.003		0.003		0.002		0.002
TELE		0.005		0.005		0.006		0.006
HUSH		0.001		0.001		0.001		0.001
OVRE	0.017	0.008		0.024	0.017	0.007		0.023
VARV		0.005	0.001	0.006		0.004		0.005
RALS		0.004		0.004		0.003		0.003
BILA	0.016	0.049	0.007	0.072	0.016	0.045	0.007	0.069
CYKL				0.001		0.001		0.001
FLYG		0.003		0.003		0.003		0.003
OVRR						0.001		0.001
INST		0.004		0.004		0.003		0.003
A_TI		0.003		0.003		0.002		0.002
EL_O		1.156	0.473	1.630		1.060	0.438	1.498
GASV		0.618		0.618		0.517		0.517
BYGG		1.599		1.599		1.467		1.467
VARU		1.572	0.085	1.657		1.430	0.078	1.508
HOTE		0.227	0.022	0.249		0.209	0.020	0.230
SAMF		4.183		4.183		3.809		3.809
POST		0.101		0.101		0.092		0.092
BANK		0.169	0.004	0.174		0.156	0.004	0.160
FAST		0.746	0.037	0.783		0.687	0.034	0.720
UPPD		0.659	0.022	0.681		0.605	0.020	0.625
REPA		0.091		0.091		0.083		0.083
OVRP	0.142	0.481		0.623	0.143	0.445		0.588
TOTAL	1.568	14.096	0.740	16.405	1.559	12.787	0.686	15.032

Table 5.13. Impacts on welfare and aggregate carbon emissions of all scenarios.

Scenario	Aggregate welfare impact		Aggregate CO_2 emissions		
	b.SEK	%	Domestic	Foreign	Global
BENCH			50029	11786	61815
C100	−3.9	−0.3	−52.2	5.6	−46.6
DIES	−0.4	0	−6	0.1	−5
PETROL	−3.0	−0.2	−42	−0.6	−42

Specifically, we need to add better data on the differences in factor endowments of households, to better reflect differences in their income sources; incorporate data-based estimates of leisure consumption in the benchmark, as well as labour supply elasticities for different household types; and employ data-based estimates of differences in carbon emissions in foreign countries relative to Sweden. As revised estimates for these parameters are generated they can be introduced directly into the existing model instantly.

The model may also be incomplete in terms of its treatment of the economic structure of some sectors. Specifically, we could provide a richer specification of the production technology of sectors with respect to energy inputs (e.g. allowing energy to be an input that combines with other intermediates in a Leontief manner, but which incorporates some degree of substitutability between energy types; and the use of non-separable production functions); incorporate some supply restrictions on imports in key sectors, particularly those such as electricity which may be constrained by resource availability and/or network logistics; model the way in which labour taxes impact households in a way that captures differences between marginal and average rates, as well as differences across households; model the effects of unemployment, including implications for unemployment benefits and the government budget; and model the use of nuclear and non-nuclear technologies more explicitly, perhaps with a formal sub-model of the electricity sector. Each of these extensions is conceptually straightforward and uses relatively familiar modelling tools, but is beyond the scope of the current project. We believe that each could be significant for current policy purposes.

The model could also be evaluated in terms of more radical changes in structure. Specifically, we could consider incorporating measures of environmental benefits[28] explicitly into the household utility function, to allow a complete cost–benefit analysis to be undertaken; explicit dynamics, with attention given to the rate at which households and firms discount future environmental benefits relative to current costs; lobbying activities surrounding green tax reforms, and endogenous political activity over the selection of reforms; and endogenous technical change induced by carbon taxes. Each of these entail exciting methodological extensions.

Notes

1. We are grateful to Åke Nordlander and Maude Svensson of the Swedish Ministry of Finance for discussions on these issues, to Runar Brannlund, Roberto Roson, Tom Rutherford and a referee for helpful suggestions, and to the Swedish Treasury Department for financial support.
2. Of the total change in tax revenues, estimated at about 90 billion SEK, energy and environmental taxes were estimated to generate 3 billion SEK in the absence of changes in the VAT treatment of energy. The addition of VAT on energy added another estimated 14 billion SEK in revenue (Åke Nordlander, personal communication).
3. The input–output database formally identifies 88 sectors, but one of these is effectively a 'dummy' sub-industry which contains no transactions and is therefore deleted. We therefore refer to the model as having 87 sectors.
4. The primary argument for aggregation, given the ready availability of powerful software and hardware for these models, has to do with the 'reliability' of data and priors at the proposed level of aggregation. Several of the data items required for our analysis are only available at an aggregated level, although far fewer than one would think and still at a relatively disaggregated level of about 20 or 30 sectors. Harrison and Kriström, (1997, Appendix A) document our data collation efforts, and the instances where we needed to map one aggregate sector into several of our disaggregated sectors. For example, basic data on factor payments were generally available only at the 3-digit SNR level, while our full model employs many 4-digit sectors. Hence we needed to use the former as the basis for individual sectors at the latter level of disaggregation. With respect to the use of *a priori* judgements, our belief is that it is much easier to apply serious priors to detailed sectors than it is to synthetic aggregates. In any event, if the priors in question are essentially held in a diffuse manner over a range of sectors, then nothing is lost if one so applies them in our disaggregated model. Providing the reader knows when such uniform assumptions are being applied, and is not dazzled by the fake detail of the analysis, it is foolish to 'hardwire' in the level of application of priors by aggregation. Formal decision-theoretic methods of aggregation of input–output sectors are explored by Harrison and Manning (1987) and provide statistically informative alternatives to naïve aggregation as practiced by many early-generation CGE modellers. However, sophisticated or naïve aggregation is simply misplaced in the present setting.
5. This formulation employs a nested production function in which K and composite labour substitute at the top level to produce value added in a given sector. At the bottom level the labour types then substitute to produce the composite labour factor. Both levels are CES, hence setting the elasticities of substitution at each level to the same value results in the nests collapsing into one level in which the three substitute at that rate.
6. It is common to assume in the short term that factors are likely to be sector-specific, and in the long term that factors tend to be mobile across sectors. We would expect a short run model of this kind to generate smaller welfare gains from a 'first-best' liberalization, since resources are constrained in their ability to reallocate to more productive uses. On the other hand, we would expect the short run model to exhibit less extreme changes in production structure since the sector-specificity of factors generates less elastic supply schedules. We also recognize that some factors are likely to be specific to one or other sectors even in the long run. An obvious example might be the natural resources used in mining.
7. Although we do not offer a detailed model of the rigidities in the oil and extraction sectors, this feature of our model is similar in effect to the model used in Bovenberg and Goulder (1995a, fn.15).
8. For those steeped in the traditions of input–output analysis.
9. Since the matter continues to be confused by commentators that should know better (e.g. Jorgenson and Wilcoxen (1995, p. 176), we stress that the assumption of a Leontief technology is not mandated by our use of the calibration approach to estimation, nor by computational constraints. In general we do restrict ourselves to nested-CES functions, although they can be

used to represent globally regular functional forms in a locally flexible manner (see Perroni and Rutherford, 1995a, 1995b).

10. The rates are defined legally as falling on the use of one of several primary energy types. We estimate the physical use of each energy type in each sector, then estimate the value of the use of each energy type in each sector by applying average 1995 prices for each type, and then infer the value of carbon (sulphur) taxes paid by each sector on its use of each energy type. We then aggregate these inferred tax payments, aggregate the payments for the use of energy by that sector, and calculate an *ad valorem* carbon tax on a net basis. These calculations allow us to generate carbon tax estimates for each sector that properly reflect the primary energy use of each sector.
11. The popular reason for using higher trade elasticities is that one can thereby avoid these effects, which are deemed unlikely *a priori* for a country as small in international trade terms as Sweden. Although the specification of trade elasticities that mitigate these effects is more involved than just assuming large or small values (see Harrison et al., 1997), these are not debates which are relevant here.
12. It should be noted that the STEN sector also has some oil importing activity, all of which is sold to the PETR sector.
13. These calculations use 1995 data.
14. The energy tax varies across different types of diesel and petrol. These numbers represented un-weighted averages across different classes of diesel and petrol. Source: Treasury of Sweden, 1996.
15. We do not distinguish children from the rest. If one wants to do so, then the use of household shares as a proxy has the unfortunate implication of unduly penalizing multiple-individual households. It would be possible to make some plausible inferences about the number of children in each of our household groups, given the way that they are defined, but we see no logic in disenfranchising those that happen to be politically disenfranchised by current voting entitlements.
16. The welfare changes are measured in terms of income-equivalents expressed in SEK/year. These income values are derived by applying the percentage change in utility to the benchmark income level of the household. If the percentage changes in utility are the same across households then richer households will have a larger income change due solely to their larger base incomes in SEK.
17. At the time of writing, by the first author.
18. These speculations are supported by inspection of the differences across household expenditure shares that are driving these results in our model, but are not modelled formally as a household technology with these scale effects.
19. The other groups, single and cohabiting households, only include one or two adult persons, respectively.
20. The term ktons refers to 1000 tons.
21. Which is sometimes stated as the importance of being unimportant, in the sense that the smaller (greater) the share of an input in cost the smaller (greater) will be the absolute value of the derived demand elasticity for the input. This law is valid in the present case, since the elasticity of product demand (around 1) clearly exceeds the elasticity of input substitution (we are referring to intermediate inputs which have an assumed elasticity of substitution of 0.25 in our model).
22. There is some controversy in international negotiation circles as to whether or not foreign-induced emissions should be counted towards a country's contributions to changes in global carbon emissions. Apart from the obvious point of avoiding double-counting, this is a non-debate: of course they should. It is another matter to debate legal liability for policing foreign economic activity induced by (internationally legal and acceptable) domestic policies (see Harrison, 1994). Our concern here is to inform the policy debate in Sweden, not to posture by generating strategically creative environmental accounts for negotiators.

23. Specifically, we assume that Japan is just as efficient (due to nuclear power use), Norway is just as efficient (due to hydro power), the European Union countries are 50% less efficient, the United States is 100% less efficient, and the rest of the world is 200% less efficient. These aggregate efficiency measures are used to scale up the sectoral emissions generated internationally by Sweden, depending on the endogenous source of imports. It should be possible to refine these estimates of foreign emissions in time.
24. There is a sixth source: wood. There were substantial intermediate sales from the SKOG sector to the EL_O sector in 1992, comparable in value to sales from the GASV sector. These inputs represent the use of wood scraps to generate supplementary electricity in some specialized pulp factories. Since it is not liable for carbon taxes, we ignore it in our discussion.
25. The current specification of technology in our model does not differentiate energy inputs from non-energy inputs. Hence the derived elasticity of demand for UPPD would be about the same as for PETR in the model, given that the intermediate input cost shares are about the same for EL_O. An extension of the model could add this differentiation, allowing an extremely low elasticity of substitution between energy and non-energy inputs as composites, but some substitution between the items within each composite. In such a version it would be harder for EL_O to substitute away from taxed inputs. The only way it could do so would be to substitute towards the EL_O energy input, which we interpret as nuclear-generation. If we further added constraints on that avenue of escaping taxes by substitution, such as specified in the N100 scenario in Harrison and Kriström (1997), the EL_O sector would be hit much harder by the carbon tax increase.
26. The source for these estimates is their summary report, available on web site http://www.unep.ch/ipcc/sumwg3.html. The estimates appear near the end of page 7 of that report.
27. We are considering here only the gross benefits from the reduction in carbon emissions that would flow from the proposed policies, since those were the ones that were claimed for them in the policy debate. Since there are other externalities associated with the use of transportation, our analysis should not be viewed as a complete cost–benefit calculation. The most significant such externalities are emissions of other pollutants, congestion and the lost time spent in transit, and the risk of accident. Small and Gómez-Ibáñez (1998) provide a good review of the literature on externalities from transportation. It is not obvious that all of these externalities are positive or negative for all households, however.
28. Some analysts have proposed using the estimated cost of the existing carbon tax structure as a crude measure of the environmental benefits from reduced emissions. We regard this inference as problematic, to put it politely. At best it could be inferred that the murky political process leading to the existing tax level represents the median voter, and then only if one were to make heroic assumptions about that political process representing the outcome as if a series of dichotomous-choice referenda had been undertaken at alternative tax-prices. Although a matter of some controversy (see Harrison and Kriström, 1995), such inferences cannot even be made if one uses a hypothetical survey to mimic the results of real referenda of this type (see Cummings et al., 1995, 1997). Even assuming away these problems, knowing the marginal value that the median voter places on some public good tells us nothing whatsoever about the distribution of benefits, at least in the absence of super-heroic assumptions that would make even Fantomen blush. Without information on that distribution one cannot say anything about the net impact on welfare of individual households, or even about the aggregate impact under simplifying utilitarian assumptions. There is simply no acceptable substitute for estimating those benefits directly, and accounting fully for the potential biases in hypothetical survey elicitation procedures (see Blackburn et al., 1994).

References

Anderson, K. and E. Norrman, 1987, Capital taxation and neutrality. a study of tax wedges with special reference to Sweden, Lund Economic Studies Number 41, Lund: University of Lund.

Bergman, L., 1991, Tillväxt och miljö, Bilaga 21 till LU-90, Stockholm: Finansdepartmentet. (In Swedish; 'Growth and the Environment', Supplement 21 to the Swedish Medium-Term Economic Survey), Stockholm: Treasury of Sweden.

Blackburn, M., G.W. Harrison and E.E. Rutström, 1994, Statistical bias functions and informative hypothetical surveys, American Journal of Agricultural Economics, 76, 1084–1088.

Bovenberg, A.L. and L.H. Goulder, 1995a, Costs of environmentally motivated taxes in the presence of other taxes: general equilibrium analyses, Working Paper No. 5117, National Bureau of Economic Research.

Bovenberg, A.L. and L.H. Goulder, 1995b, Optimal environmental taxation in the presence of other taxes: general equilibrium analyses, Unpublished Manuscript, Stanford: Stanford University.

Brooke, A., D. Kendrick and A. Meeraus, 1992, GAMS: A User's Guide, Release 2.25, Danvers, MA: Boyd & Fraser.

Cummings, R.G., S. Elliott, G.W. Harrison, and J. Murphy, 1997, Are hypothetical referenda incentive compatible? Journal of Political Economy, 105, 609–621.

Cummings, R.G., G.W. Harrison and E.E. Rutström, 1995, Homegrown values and hypothetical surveys: is the dichotomous choice approach incentive compatible?, American Economic Review, 85, 260–266.

de Melo, J, and D. Tarr, 1992, General Equilibrium Analysis of US Foreign Trade Policy, Cambridge, MA: MIT Press.

Dirkse, S.P. and M.C. Ferris, 1995, The PATH solver: a non-monotone stabilization scheme for mixed complementarity problems, Optimization Methods and Software, 5, 123–156.

Goulder, L.H., 1995a, Environmental taxation and the double dividend: a reader's guide, International Tax and Public Finance, 2, 157–183.

Goulder, L.H., 1995b, Effects of carbon taxes in an economy with prior tax distortions: an intertemporal general equilibrium analysis, Journal of Environmental Economics and Management, 29, 271–297.

Harrison, G.W., 1994, Environmentally sensitive industries and an emerging Mexico, North American Journal of Economics and Finance, 4, 109–126.

Harrison, G.W. and B. Kriström, 1995, On the interpretation of responses in contingent valuation surveys, in P.-O. Johansson, B. Kriström and K.-G. Mäler, eds., Current Issues in Environmental Economics, New York: Manchester University Press.

Harrison, G.W. and B. Kriström, 1997, Carbon taxes in Sweden, Final Report, Stockholm: Skatteväxlingskommittén.

Harrison, G.W. and R. Manning, 1987, Best approximate aggregation of input–output systems, Journal of the American Statistical Association, 83, 1027–1031.

Harrison, G.W. and H.D. Vinod, 1992, The sensitivity analysis of applied general equilibrium models: completely randomized factorial sampling designs, The Review of Economics and Statistics, 74, 357–362.

Harrison, G.W., T.F. Rutherford and D.G. Tarr, 1993a, Piecemeal trade reform in the partially liberalized economy of Turkey, World Bank Economic Review, 7, 191–217.

Harrison, G.W., R. Jones, L.J. Kimbell and R. Wigle, 1993b, How robust is applied general equilibrium analysis? Journal of Policy Modeling, 15, 99–115.

Harrison, G.W., T.F. Rutherford and D.G. Tarr, 1995, Quantifying the outcome of the Uruguay round, Finance & Development, 32, 38–41.

Harrison, G.W., T.F. Rutherford and D.G. Tarr, 1996, Increased competition and completion of the market in the European Union: static and steady state effects, Journal of Economic Integration, 11, 332–365.

Harrison, G.W., T.F. Rutherford and D.G. Tarr, 1997, Quantifying the Uruguay round, Economic Journal, 107, 1405–1430.

Jorgenson, D.W. and P.J. Wilcoxen, 1995, Intertemporal equilibrium modeling of energy and environmental policies, in P.-O. Johansson, B. Kriström and K.-G. Mäler, eds., Current Issues in Environmental Economics, New York: Manchester University Press.

Nordhaus, W.D., 1995, The Swedish dilemma: nuclear energy v. the environment, Report to Studieförbundet Näringsliv och Samhälle, Stockholm.
Perroni, C. and T.F. Rutherford, 1995a, Regular flexibility of nested CES functions, European Economic Review, 39, 335–343.
Perroni, C. and T.F. Rutherford, 1995b, A comparison of the performance of flexible functional forms for use in applied general equilibrium analysis, Discussion Paper 95–6, Colorado: University of Colorado.
Reinert, K.A. and D.W. Roland-Holst, 1992, Armington elasticities for the manufacturing sectors of the United States, Journal of Policy Modeling, 14, 631–639.
Reinert, K.A. and C.R. Shiells, 1991, Trade substitution elasticities for analysis of a North American free trade area, Unpublished Manuscript, Washington, DC: Office of Economics, US International Trade Commission.
Rutherford, T.R., 1992, Applied general equilibrium modeling with MPSGE as a GAMS subsystem, Economics Working Paper 92-15, Colorado: University of Colorado.
Rutherford, T.F., 1993, MILES: a mixed inequality and nonlinear equation solver, Economics Working Paper, Colorado: University of Colorado.
Rutherford, T.F., 1995, Extensions of GAMS for complementarity and variational problems arising in applied economics, Journal of Economic Dynamics and Control, 19, 1299–1324.
Rutherford, T.F., E.E. Rutström and D.G. Tarr, 1994, L'Accord de libre echange entre le Maroc et la CEE: une evaluation quantitative, Revue d'Economie du Developpement, 2, 97–133.
Small, K.A. and J.A. Gómez-Ibáñez, 1998, Urban transportation, in P. Cheshire and E.S. Mills, eds., Handbook of Regional and Urban Economics, Volume 3: Applied Urban Economics, Amsterdam: North-Holland, 1998 forthcoming.
Statens Offentliga Utredningar (SOU 1990: 59), 1990, Sätt Värde på Miljön: Miljöavgifter och andra Ekonomiska Styrmedel. (In Swedish: Final Report from the Commission on Environmental Charges). Stockholm: Norstedts.
Statens Offentliga Utredningar (SOU 1994: 85), 1994, Ny Lag om Skatt på Energi: En Teknisk Översyn och EG-anpassning. (In Swedish: Report from the Committee on Energy Taxation, Part 1). Stockholm: Norstedts.
Statens Offentliga Utredningar (SOU 1995: 64), 1996, Klimatförendringar i trafikpolitiken. (In Swedish with English Summary: Climate Change in Traffic Policy, Report from the Committee on Traffic and Climate). Stockholm: Norstedts.
Statens Offentliga Utredningar (SOU 1996: 26), Ny Kurs i Trafikpolitiken. (In Swedish with English Summary: New Directions in Traffic Policy, Report from the Commission on Communications). Stockholm: Fritzes.
Treasury of Sweden, 1995, Energi och Miljörelaterade Skatter I Sverige och I OCED-länderna. (In Swedish: Energy and Environmentally Related Taxes in Sweden and the OECD Countries), Unpublished manuscript, Stockholm: Treasury of Sweden.

CHAPTER 6

Evaluating External Costs and Benefits Resulting from a Cleaner Environment in a Stylized CGE Model[1]

Giancarlo Pireddu

6.1. Introduction

Production and consumption activities cause a wide range of damage. Such damage is called external costs when they are not reflected in the market prices of the products. There are different types of external costs: for example, transport is responsible not only for air pollution but also for accidents, health diseases, traffic congestion, noise, barrier effects and visual intrusion. Estimating the costs associated with such damage is extremely difficult. The state of the art of environmental evaluation is poor due to the lack of specific studies of the relationship between pollutants and their impact on the ecosystem (for Italian data see Pavan, 1995).

The need for better data and information regarding the environmental impacts of different pollutants generated by the fuel cycles has motivated the 'ExternE Project' (European Commission DGXII, 1994), a collaborative study between the European Commission and the US DOE. The crucial phase of the ExternE Project is represented by a systematic approach to the evaluation of impact pathways and external costs of a wide range of different fuel cycles.[2]

The 'Auto/Oil Programme' (European Commission DGIII, 1995) also aims to provide policy makers with an objective assessment of possible measures that could be adopted within the road transport sector in order to achieve higher standards of air quality.[3]

Such research efforts promise to deliver a large amount of data which will prove useful in characterizing interactions between the economic and the ecological systems. It is, therefore, necessary to be able to use a modelling framework which can take into account those interactions. Such a modelling framework should be based on general, rather than partial, equilibrium considerations which are necessary for a proper comparison of costs and benefits of environmental policy.

R. Roson and K.A. Small (eds.), Environment and Transport in Economic Modelling, 118–151.

Our purpose is to make future empirical analysis and comparisons of the potential of ecological tax reforms. The aim of this paper is to build a simple general equilibrium model whose specific objectives are:

1. to test the suitability of dose–response functions in standard production and utility functions;
2. to provide estimates of external costs to varying levels of pollution;
3. to provide estimates of benefits from a cleaner environment;
4. to describe pre-existing tax distortions, to examine various fiscal reform options considering both the tax interaction effect and the revenue-recycling effect;
5. to provide insight into the interactions among pollution levels and emission abatement technologies, fiscal instruments and economic agent behaviour.

A computable general equilibrium (CGE) model of a competitive stylized economy is developed to test the feasibility of the above objectives. In particular our CGE model considers producer–producer externalities (i.e. the damage caused to the victims' output) and producer–consumer externalities (i.e. the damage caused to the victims' utility as a result of the same pollution).

Different simulations are run. We compare economic distortions generated by traditional (i.e. non-environmental) taxes in the absence of any type of negative externalities with economic distortions generated by negative externalities. Results will show that negative externalities can generate considerable additional distortions in labour and commodity markets.

We assess two environmental policies: the unilateral abatement of emissions by the polluter (the so-called voluntary or altruistic policy) and the abatement of emissions to a level corresponding to that achievable by an ecological tax that equals the value of marginal damage caused by polluter (the so-called Pigouvian policy). Numerical results will demonstrate the superiority of Pigouvian policy.

We also assess differential incidence tests. We suppose the Government adopts a fiscally neutral stance on the Pigouvian tax, using revenues from the Pigouvian taxation to finance reductions in incentive-distorting taxes levied on labour inputs or output of the polluting sector. This revenue-neutral tax switching provides both environmental improvements and reduction in distortionary costs of the fiscal system, generating a double dividend in the strong form, according to Goulder's definition.[4] Hence the Pigouvian tax cuts back, rather than imposing, excess burden generated by distortionary taxes, and provides substantial environmental improvements.

Simulation results can demonstrate this result as external diseconomies are explicitly considered in our CGE model. A measure of the excess burden generated from this tax switching policy is computed in terms of marginal cost of public funds (MCF). The MCF test will show that households take advantage of the Pigouvian policy because swapping the Pigouvian tax for a distortionary tax causes positive net benefits.[5]

Another important finding of our exercises is that any environmental policy that neutralises the effects of negative externalities seems to increase total Government revenue as a general increase in the economywide production also generates a permanent tax base increase. A permanent reduction in a distortionary tax revenue can be provided by real effects of a cleaner environment. We have found this result in a static approach and in the future we hope to show it in a dynamic CGE approach.

6.2. Basic features of the CGE model

6.2.1. General specifications

This section describes a simple static CGE model of a competitive economic system used to perform counterfactual analysis. The basic structure of the model is borrowed from standard models currently in use, described in Shoven and Whalley (1992). The way in which this model differs from the standard ones is the inclusion of negative technical externalities in the production function, defensive expenditure in the utility function, and an 'end-of-pipe' pollution abatement cost function.

Negative technical externalities cause physical and economic losses in production, consumption and tax revenue. In the case of production, the victim obtains less output from a given set of inputs as it can not mitigate the physical damage. For example, emissions of SO_2 and ozone cause crop losses of corn, soybeans and wheat.[6]

Defensive expenditure arises when physical damage can be mitigated. For example, traffic noise can be mitigated by installing double glazing. In this case defensive expenditure takes the form of additional purchases of goods for a consumer.

We assume that the pollution abatement cost using an end-of-pipe technology takes the form of purchases of additional inputs (labour, capital and intermediate inputs) for the polluter.

Producer behaviour

The model is disaggregated into three activity sectors, in each of which there is a representative profit maximizing firm, with constant returns to scale production technology. There are two homogeneous primary factors of production, labour and capital, each of which is mobile across sectors. Capital and labour endowments are owned by two representative consumer groups. Labour endowment is equal to labour supplies plus the consumption of leisure.

Capital and labour produce value added according to the Cobb–Douglas (CD) function. The model uses a 3×3 fixed-coefficient input output matrix A, with columns giving the intermediate inputs required for each unit of output. Producer behaviour is characterized by minimizing costs for each unit of output. As a result of our assumption of perfect competition, all of the firms, after making payments for factors, intermediates and taxes, make zero profit.

Consumer behaviour

The consumer's decision problem is simplified in this model due to the absence of intertemporal decisions ('no saving' hypothesis). Therefore, it is assumed that all the consumers' income is spent. A constant elasticity substitution (CES) function determines the consumer's choice between the consumption of leisure and an aggregate good. A subsequent nested step (which is of the CD type) determines the division of aggregate good consumption among consumption goods.

Government

The government plays a crucial role. It collects tax revenues, as well as it purchases goods and redistributes revenues in the form of transfers to households. We assume no deficit or surplus in the government balance. In our model it is also assumed that consumer choices are independent of government expenditure. In reality this is not true because consumers take advantage of government services. In other words, the household's utility generated from private goods may be correlated to the government expenditures.

6.2.2. *Victims' behaviour*

Producer's behaviour

Ballard and Medema (1993) model the producer–producer externality in terms of defensive expenditures, i.e. the victim must use more intermediate inputs than they would otherwise have to use. In this paper we use a damage (dose–response) function.

Let Q_{S2} be the polluter's output, let Q_{S1} be the victim's output, let $\bar{\gamma}$ be the (fixed) polluting coefficient per unit of polluting output, and let x be the unity abatement share of the polluting coefficient (i.e. $0 \leq x \leq \bar{\gamma}$). Our general form of the unity damage function d between two producers[7] (i.e. victim and polluter) is given by

$$d(Q_{S2}, \bar{\gamma}, x) = z(1 - e^{-\phi Q_{S2} \cdot (\bar{\gamma} - x)^{\theta}}) \quad 0 \leq d \leq z \tag{6.1}$$

where ϕ and θ are parameters that show the amount by which the net pollution damages the victim's production, and z is an asymptotic parameter which can not be greater than one. The value of the parameter ϕ representing the scale of damage is obtained by calibration, as is customary in CGE models. The parameter θ is 'guesstimated' and, according to convexity assumption of the conventional theory, it must be >1 to satisfy both first and second-order conditions (details in Appendix). The parameter θ represents the non-linearity in the damage function.[8]

The unity damage function guarantees that the level of d will be between

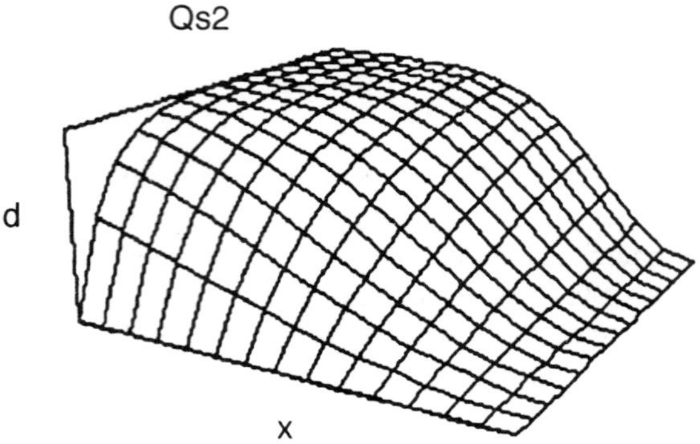

Figure 6.1. Unity damage, abatement ratio and polluting output.

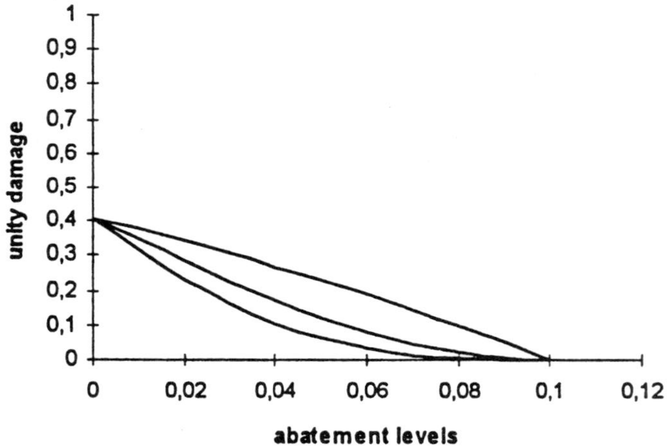

Figure 6.2. Damage and unity abatement levels.

zero and z. This structure captures the fact that the damage suffered by the victim will change according to variations in both the polluter's output and the net pollution level (Figure 6.1).

Figure 6.2 shows the unity damage level, for a given level of polluting output and pollution coefficient, in function of increasing abatement level. Three curves are shown for different values of θ whose role is crucial in the dose–response function. θ is set equal to 3 in the bottom curve, to 2 in the middle curve, and to 1 in the top curve. The damage vanishes when the unity abatement level is equal to the fixed pollution coefficient $\bar{\gamma}$ which is set equal to 0.1 in the simulations.

Figure 6.3 shows unity damage levels in function of increasing polluter's output for a given unity abatement level. Three curves are again shown for different values of θ. θ is set equal to 3 in the right curve, to 2 in the middle

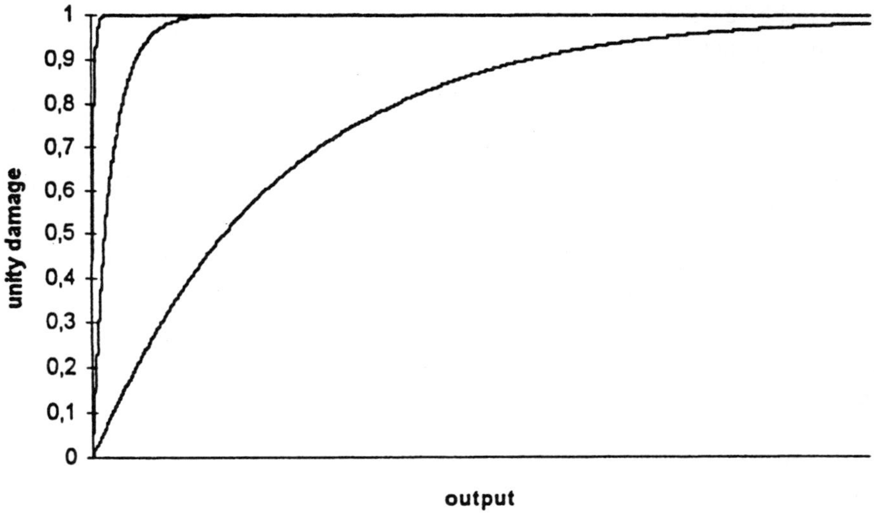

Figure 6.3. Damage and polluter's output levels.

curve, and to 1 in left curve. The unity damage can not be greater than the asymptotic level, which is set equal to 1 in the figure.

Other functional forms, including logistic and Gompertz functions, could be considered (see Conrad and Clark, 1987, Ch. 2).

If the production function is assumed to be of Cobb–Douglas type (constant returns to scale), then the victim's production function becomes

$$Q_{S1} = (1 - d) \cdot cL^{\alpha} K^{1-\alpha} \tag{6.2}$$

where c is the parameter which scales the production and α the parameter which weights labour L and capital K inputs.

In the presence of negative externalities, less output is produced by primary factors. In the functional form used in this paper, intermediate inputs are assumed to be used in fixed proportions as a function of the level of output, as described by the Leontief hypothesis. We assume further that the optimal combination between labour and capital inputs depends only on the relative factor price ratio. In other words negative externalities have no effect on the optimal labour/capital ratio.

The victim's external cost function, EC, of the current damage rate is given by

$$EC_{Q_{S1}} = P_{S1} \cdot d \cdot (cL^{\alpha} K^{(1-\alpha)}) \tag{6.3}$$

where P_{S1} is the victim's producer price.

We assume that the total external damage cost is a positive convex function of the level of pollution, or a negative convex function of the level of polluting abatement (details in Appendix 1).

Consumer's behaviour

This model includes a producer–consumer externality. We suppose there are two representative consumers. The h_{th} ($h = 1, 2$) consumer has to decide how to split the consumption across consumption of leisure time L_h, composite commodity CON_h and composite defensive commodity $DEFC_h$ which is required to cope with the negative externality. Demand of defensive expenditure depends on the current net pollution level $(\bar{y} - x)$.

Current consumption is modelled by the h_{th} consumer undertaking a two stage optimisation. In the first stage the h_{th} consumer starts with a budget Y_h that equals the rental value of capital endowment and labour endowment (whether sold or retained as leisure), plus government transfers, minus taxes. Y_h is not the observed money income but the so-called 'expanded' income. The implicit price of the composite commodity, CON_h, is $PAGGC_h$ and of the composite defensive commodity, $DEFC_h$, is $PAGGD_h$. The price of leisure, P_{lh}, is set equal to the wage rate net of consumer's personal marginal income tax rate, PIT_h, i.e. $P_{lh} = w(1 - PIT_h)$.

The behaviour function of the h_{th} consumer is given by

$$\begin{aligned} &U_h(CON_h, L_h, DEFC_h(\bar{y}, x)) \\ &= [\beta_{0h}^{1/\sigma} CON_h^{(\sigma-1)/\sigma} + \beta_{1h}^{1/\sigma} L_h^{(\sigma-1)/\sigma} \\ &\quad + (1 - \beta_{0h} - \beta_{1h})^{1/\sigma} \cdot (DEFC_h \cdot (\bar{y} - x)^{\tau})^{(\sigma-1)/\sigma}]^{\sigma/(\sigma-1)} \end{aligned} \tag{6.4}$$

where β's are weighting parameters, τ is a damage parameter, and σ is the elasticity of substitution among L_h, CON_h and $DEFC_h$. All parameters vary according to the households. (the symbol h is omitted for simplicity in the case of the elasticity of substitution parameter).

Maximizing the utility function subject to the budget constraint gives the conditional demands. They are given by

$$CON_h = \beta_{0h} Y_h / (PAGGC_h^{\sigma} \cdot \Delta_h) \tag{6.5}$$

$$L_h = \beta_{1h} Y_h / (P_{1h}^{\sigma} \cdot \Delta_h) \tag{6.6}$$

$$DEFC_h = (1 - \beta_{0h} - \beta_{1h})(\bar{y} - x)^{[\tau(\sigma-1)+1]/\tau} \cdot Y_h / (PAGGD_h^{\sigma} \cdot \Delta_h) \tag{6.7}$$

where

$$\begin{aligned} \Delta_h = &PAGGC_h^{(1-\sigma)} \beta_{0h} + P_{lh}^{(1-\sigma)} \beta_{1h} \\ &+ PAGGD_h^{(1-\sigma)} (1 - \beta_{0h} - \beta_{1h})(\bar{y} - x)^{[\tau(\sigma-1)+1]/\tau} \end{aligned}$$

and P_{1h}, $PAGGC_h$ and $PAGGD_h$ are implicit prices of the leisure time and the two composite commodities. Each h_{th} consumer has its own implicit price.

In the second stage, we assume that the h_{th} consumer minimises the cost of acquiring the specific commodities which compose both composite good CON_h and $DEFC_h$, according to a Cobb–Douglas function. Given the desired level of composite good CON and DEFC, 'standard' goods conditional demands

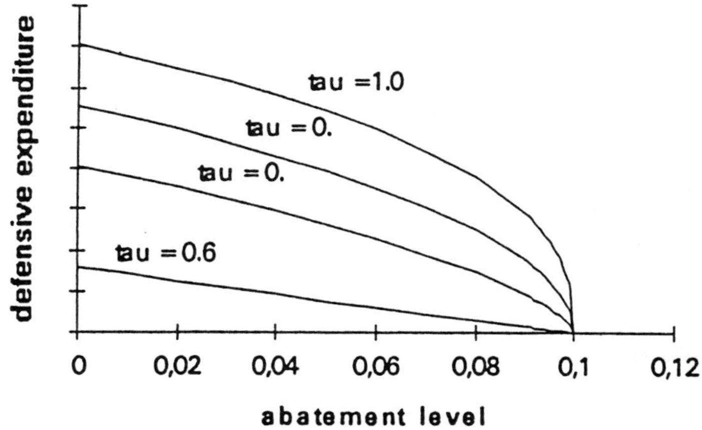

Figure 6.4. Defensive expenditure and pollution abatement.

C_h and 'defensive' goods conditional demand CD_h are respectively given by

$$C_{ih} = \lambda_{ih} \frac{PAGGC_h \cdot CON_h}{MP_i} \quad i = \text{S1, S2, S3} \tag{6.8}$$

$$CD_{ih} = e_{ih} \frac{PAGGD_h \cdot DEFC_h}{MP_i} \tag{6.9}$$

where MP_i is the market price of the i_{th} commodity, i.e. $MP_i = P_i(1 + tax_i)(1 + vat_i)$, and λ_i and ε_i are expenditure share parameters that vary according to the households. Indirect tax rates are excise duty (*tax*) and value added taxes (*vat*).

We also assume that defensive consumption can be used as a proxy of the damage value suffered by households in the absence of specific (i.e. health, food, natural or man-made impacts) marginal external costs. The h_{th} consumer's total external cost of the current emission abatement rate is given by

$$EC_h = PAGGD_h \cdot DEFC_h(x) \tag{6.10}$$

Defensive expenditure is increasing in unity net pollution level. We assume that total expenditure function (eqn. 6.7) is a positive concave function of the level of pollution, and a negative concave function of the level of abatement (see details in Appendix 2). The relationship between pollution abatement level and defensive expenditure is shown in Figure 6.4, which shows, for a given parameter σ, different defensive expenditure curves for increasing value of τ exponent.

The 'green' or restricted (concave) utility function, given by

$$U_h = U_h(CON_h, L_h), \tag{6.11}$$

is defined only in terms of standard consumption CON_h and leisure L_h.

Finally, the reader will realise that household's utility functions (eqns 6.4, 6.11) are independent of government expenditure. In other words it is supposed, as is customary in CGE models, that public expenditure has no effect on ordinary demand function. Moreover it is evident that the provision of public goods by the government contributes to an increase in the household's private welfare. This issue is not tackled in this paper (see Rossier, 1976; Aprile, 1984). The question of whether the efficiency effects of ecological taxes are confirmed if the equity effects, i.e. the role of income distribution,[9] is not considered in this paper (see Baumol, 1974).

6.2.3. *Polluter's behaviour*

Abatement cost function

The total cost depends on both the polluter's output and the emission reductions. In our model we suppose that the total abatement cost is a positive convex function of the level of pollution abatement, and by implication is a negative convex function of the level of pollution up to some level where total abatement cost is zero.

These desired properties suggest a function of the form

$$TAC(Q_{S2}, x) = vQ_{S2}x^{\mu} \tag{6.12}$$

where v is a monetary parameter, μ is a technological parameter, and x is the unity pollution abatement and is the same included in the victim's damage function and household's defensive expenditure. These two parameters show the amount by which the pollution abatement contributes to raise the polluter's abatement cost. According to convexity assumption of the conventional theory, the parameter μ must be greater than 1 to satisfy both the first and the second order conditions:

$$\frac{\partial TAC}{\partial x} = \mu v Q_{S2} x^{\mu - 1} > 0 \quad \text{and} \quad \frac{\partial^2 TAC}{\partial x^2} = \mu(\mu - 1) v Q_{S2} x^{\mu - 2} > 0.$$

Figure 6.5 shows total abatement cost curves for different μ values.

A trade-off between inputs used for output production and inputs used for abatement production is not allowed in this model. In other words, when the polluter faces defensive expenditure, inputs demanded for the abatement technology are supposed to be additive and independent of the inputs demanded for the production of output.[10]

In this way we assume that the pollution abatement technology is separate and distinct from the technology to produce traditional goods and services. In other words, this is equivalent to assuming an end-of-pipe emission control. Only a dynamic version of this CGE model can allow to model technical progress and, hence, reductions in the pollution coefficient $\bar{\gamma}$.

This approach also assumes that extra expenses of labour, capital and intermediate inputs can be evaluated, at first glance, according to a 'Leontief'

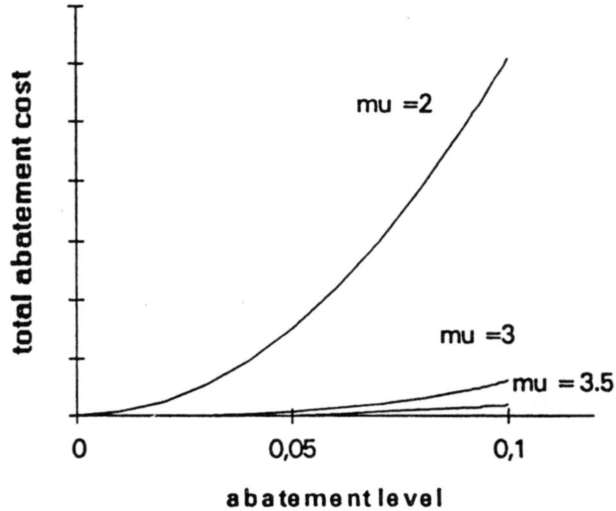

Figure 6.5. Total abatement cost.

type cost function. Consequently extra expenses of labour, capital and intermediate inputs are assumed to be in fixed proportions with the total cost, i.e.:

$$\begin{aligned} &P_L \cdot L = TAC \cdot a^d_{LS2} \quad MP_i \cdot Q_{iS2} = TAC \cdot a^d_{iS2} \\ &P_K \cdot K = TAC \cdot a^d_{KS2} \quad a^d_{LS2} + a^d_{KS2} + \sum_i a^d_{iS2} = 1 \end{aligned} \tag{6.13}$$

where $a^d_{.S2}$ are fixed abatement cost coefficients, P_L is equal to the wage rate augmented by social security taxes, i.e. $P_L = w(1 + ssc)$, and P_K is the rental price of capital services augmented by corporate taxes, i.e. $P_K = r(1 + corp)$ and MP_i is the market price of intermediate inputs.

Polluter's profit function

The polluter's profit function, Π, is subject to technology constraints, i.e. production technology and abatement technology. When the polluter faces a Pigouvian tax, the polluter's maximizing profit function (according to Ballard and Medema, 1993) is given by

$$\begin{aligned} &\max \Pi_{Q_{S2}}(L_{S2}, K_{S2}, x) \\ &= \left\{ P_{S2}Q_{S2}(L_{S2}, K_{S2}) - P_L L_{S2} - P_K K_{S2} - Q_{S2}(L_{S2}, K_{S2}) \sum_i a_{j,Q_{S2}} \cdot P_i \right\} \\ &\quad - [tQ_{S2}(L_{S2}, K_{S2}(\bar{\gamma} - x) + TAC(Q_{S2}, x)] \end{aligned} \tag{6.14}$$

where $P_{S2}Q_{S2}(L_{S2}, K_{S2})$ are revenues, $P_L L_{S2}$ and $P_K K_{S2}$ are labour and capital expenses, $Q_{S2}(L_{S2}, K_{S2}) \cdot \sum_i a_{i,Q} \cdot P_i$ are intermediate expenses of i_{th} commodity type, t is the unity Pigouvian tax and *TAC*(Q, x) is the total abatement cost function. Addends within braces determine the polluter's gross profit while addends within square brackets show, respectively, the Pigouvian tax and the abatement cost.

A Pigouvian tax is a tax on net emissions. The polluter can avoid this tax either by reducing its output or by abating emissions. The first-order condition for x, the change in net pollution per unit of output, gives

$$t \cdot Q_{S2}(L, K) = \frac{\partial TAC(x)}{\partial x} \tag{6.15}$$

i.e. the marginal benefit to the polluter from abatement is tQ, and it must be set equal to the marginal abatement cost.

The first-order conditions for L and K give, as it is customary, the conditional demands of labour and capital.

6.2.4. *Socially optimal pollution*

In the case of the Pigouvian policy the CGE model endogenously computes the optimal abatement level of emissions. Let SC be the social cost function obtained adding abatement costs and external costs suffered by both victims together. The optimal level of abatement (pollution) socially desirable is given by

$$\min SC(x) = TAC(x) + P_{S1} \cdot d(x) \cdot (cL^{a} K^{(1-a)}) + \sum_{h} PAGGD_h \cdot DEFC_h(x) \tag{6.16}$$

Differentiation of (eqn. 12) with respect to the abatement variable x gives the first-order condition:

$$\frac{\partial TAC(x)}{\partial x} = P_{S1} \cdot (cL^{\alpha} K^{(1-\alpha)}) \frac{\partial d(x)}{\partial x} + \sum_{h} PAGGD_h \cdot \frac{\partial DEFD_h(x)}{\partial x} \tag{6.17}$$

where

$$\frac{\partial d(x)}{\partial x} < 0 \quad \text{and} \quad \frac{\partial DEFD_h(x)}{\partial x} < 0$$

because of the negative externality (both first partial derivatives are shown, respectively, in Appendices 1 and 2). The first-order condition states that the marginal abatement cost must be set equal to the sum of the marginal external costs.

6.2.5. *General equilibrium*

The solution of the model is a general equilibrium in which supplies balance demands in all markets. The solution requires that supply equals demand for labour, capital and all produced goods, that the total government's tax revenue equals its spending, and that marginal abatement costs equal marginal external costs. These conditions are achieved by means of adjustments made to producer prices, factor prices and the abatement level.

Since supply and demand functions are homogeneous of degree zero in prices and income, the solution determines relative rather than absolute prices. Thus, a price must be chosen as numeraire. In this model the *numeraire* is set equal to one and is given by the index of producer prices:

$$\bar{P} = \prod_{i=1}^{3} P_i^{\delta_i} \quad \text{with} \quad \sum_{i=1}^{3} \delta_i = 1 \tag{6.18}$$

Consequently, one of the clearing conditions of the market can be suppressed because of Walras' law.

6.2.6. *Data and welfare evaluation*

The reference economy

It is convenient to separate the environmental and non-environmental impacts of simulated policies. The reference economy equilibrium is calculated in the absence of any type of taxes and any type of negative externalities. The flows of resources between the agents of our stylized economy are presented in the social accounting matrix (SAM). The SAM in Table 6.1 shows the reference economy without fiscality and externality. A unity convention is adopted, so that all prices in the equilibrium are set equal to one.

The SAM in Table 6.2 shows the new equilibrium generated by traditional (non-environmental) fiscality as computed by the CGE model. Direct and indirect taxes levied at various stages of production and consumption are correctly included (in italics) according to the tax-payer principle.

Different taxes generate different economic distortions and excess tax burdens. This static model can show two main distortions (Goulder, 1994): the labour–leisure margin and the margin of inter-commodity choice. These margins are generated in the labour and commodity markets respectively. The choice between working and enjoying leisure is made in the labour market, while the allocation of consumption expenditure across different goods is found in the commodity markets. Personal income tax distorts the labour–leisure margin by driving a fiscal wedge between the marginal social value of labour (i.e. before-tax real wage), and the marginal private value of labour (i.e. after-tax real wage). The inter-commodity margin is altered by any tax that directly affects the relative gross-of-tax prices of the commodities.

Tax rates levied are supposed to be as follows:

- value added tax (*vat*) = 0.19,
- excise duty on output (*tax*) = 0.05,
- marginal personal income tax rate (PIT): 0.1 for the first income block and 0.25 for the second,

Table 6.1A. Social accounting matrix – economy without fiscality.

Quantity	S1	S2	S3	Labour	Capital	HOUS1	HOUS2	Government	Total	Price
S1	10	50	20			40	13		133	1.000
S2	34	65	10			81	115		305	1.000
S3	19	43	5			7	24		98	1.000
Labour	27	59	24			21	74		205	1.000
Capital	43	88	39						170	1.000
				Leisure price		1.000	1.000			

Table 6.1B. Social accounting matrix – economy without fiscality values.

	S1	S2	S3	Labour	Capital	HOUS1	HOUS2	Government	Total
S1	10	50	20			40	13		133
S2	34	65	10			81	115		305
S3	19	43	5			7	24		98
Labour	27	59	24			21	74		205
Capital	43	88	39						170
HOUS1				98	51				149
HOUS2				107	119				226
Government									
Social security contribution									
Excise duties									
S1									
S2									
S3									
Value added tax									
S1									
S2									
S3									
Corporate tax									
Personal Income Tax									
Total	133	305	98	205	170	149	226		

Table 6.2A. Social accounting matrix – economy without externalities.

Quantity	S1	S2	S3	Labour	Capital	HOUS1	HOUS2	Government	Total	Price
S1	9.71	48.95	19.53			26.00	8.37	16.57	129.13	0.999
S2	33.01	63.63	9.77			52.42	73.73	66.01	298.57	1.004
S3	18.45	42.09	4.88			4.60	15.62	10.07	95.72	0.989
Labour	25.25	55.69	22.58			18.04	83.44		205.00	0.826
Capital	42.74	88.27	38.99						170.00	0.667
				Leisure price		0.620	0.620			

Table 6.2B. Social accounting matrix – economy without externalities.

Values	S1	S2	S3	Labour	Capital	HOUS1	HOUS2	Government	Total
S1	9.70	48.90	19.52			25.97	8.36	16.56	129.01
S2	33.14	63.88	9.81			52.63	74.02	66.28	299.77
S3	18.24	41.63	4.83			4.55	15.45	9.96	94.67
Labour	20.86	46.00	18.65			11.19	51.73		148.43
Capital	28.51	58.88	26.00						113.39
HOUS1				71.04	34.02			20.00	125.06
HOUS2				77.38	79.37			30.00	186.76
Government									142.79
Social security contribution	4.17	9.21	3.73						17.11
Excise duties									
S1	0.49	2.45	0.98			1.30	0.42		5.62
S2	1.66	3.19	0.49			2.63	3.70		11.67
S3	0.91	2.08	0.24			0.23	0.77		4.23
Value added tax									
S1						5.19	1.67		6.85
S2						10.50	14.77		25.27
S3						0.91	3.08		3.99
Corporate tax	11.40	23.54	10.39						45.33
Personal Income Tax						9.97	12.75		22.72
Total*	129.08	299.75	94.63	148.43	113.39	125.06	186.72	142.79	

*Numbers may vary due to rounding.

- mandatory social security cost on labour input (ssc) = 0.2,
- corporate tax rate ($corp$) = 0.4.
- lump sum transfers to household 1 ($TRANSF1$) = 20.
- lump sum transfers to household 2 ($TRANSF2$) = 30.

Parameters and model calibration

The various elasticities have been 'guesstimated' and the model has been calibrated by selecting the parameters and input–output coefficients that replicate the reference economy. We assume that consumers maximise utility and producers maximize profits. For the abatement cost function (eqn. 6.9), the selected value for μ is 2 and the appropriate value for the scale parameter v is calculated from known data on costs, output quantities and the abatement level.

For the damage function (eqn. 6.2) the selected value for θ is 2 and the appropriate value for ϕ is calculated from the known variables of the damage function (i.e. the pollution coefficient γ, abatement variable x and the output quantities for both the victim and the polluter). For the utility function (eqn. 6.4) the selected value for τ is 0.8 and σ is 0.4 for the first consumer, and 0.8 for the second. The appropriate values for β are calibrated in the normal manner.

Once the parameters of the utility and production functions have been selected the model is ready to be used to perform simulations.

Welfare evaluation

The change in consumer welfare is expressed in terms of equivalent variation EV_h, which is given by

$$EV_h = E_h(U^n, P^0) - E_h(U^0, P^0) \tag{6.19}$$

where $E(U^n, P^0)$ is the expenditure function of the h_{th} household, i.e. the expenditure necessary to achieve the new utility level U^n with old prices P^0, and $E(U^0, P^0)$ is the expenditure function of the benchmark equilibrium.

The utility level in the expenditure function is given by the 'green' utility function, i.e. equation (6.11). According to the definition of the expenditure function, adding EV to the consumer's benchmark income will result in the same change in utility that which would be arrive as result of any change in environmental policy.

The model also calculates the marginal cost of public funds (MCF) in the so-called 'differential incidence' hypothesis. In this case, when the Pigouvian tax is levied some other taxes can be reduced. The MCF gives an exact measure of the excess burden of differential type. The MCF is given by

$$MCF = -\sum_h EV_h/\Delta G \tag{6.20}$$

where ΔG is the change in government revenues.

The household's equivalent variation shown in equation (6.19) is supposed to be independent of government expenditure GE. In our model it is assumed that the government expenditure level is not fixed but can vary in order to guarantee government balance equilibrium. As total tax revenue does vary according to alternative policies considered in our simulations, also consumers' welfare does vary according to alternative recycling tax policies. In an attempt to capture the trade-off between private and public expenditure, we assume that it is possible to compute a total equivalent variation (TEV) by summing the household's EV_h to 'government equivalent variation' EV_G:

$$TEV = -\left(\sum_{h=1}^{2} EV_h + EV_g\right) \Big/ \Delta G \tag{6.20}$$

where

$$EV_G = \sum_i P_i^0 \cdot (GE_i^n - GE_i^0) \quad i = S1, S2, S3$$

and GE_i^0 is the sectoral government expenditure (quantity) in the benchmark and GE_i^n is the new one.

6.3. Simulations and results

6.3.1. Negative externality in the reference economy

The benchmark equilibrium used for comparisons is shown in Table 6.3. It shows the additional distortions in labour and commodity markets generated by negative externalities with respect to the reference economy shown in Table 6.2. Pollution, generated by the activity sector S2, directly reduces the output of the activity sector S1 and affects the two representative households. We assume that the producer–victim suffers a reduction in labour and capital productivity as less output is obtained from any given set of inputs. We assume that households have to purchase defensive goods (and eventually leisure, although this case is not considered in this paper) in order to protect their environmental quality levels. Box 6.1 and box 6.2 present the equations and the variables that generate the equilibrium in Table 6.3.

External costs are not internalized by the polluting firm. Consequently the polluter's price does not correctly reflect the opportunity cost imposed on the victims. This results in over production and sales of the polluting goods. Purchases of defensive commodities are not included in the utility index. In other words defensive expenditures incurred to defend households from the adverse side effects of growing pollution are subtracted from households' total consumption.

Technical negative externalities reduce, in more or less extent, sectoral outputs and, consequently, the tax base. Hence, total Government revenue is reduced from 142.79 to 118.43

Table 6.3A. Social accounting matrix – economy with negative externalities.

Quantity	S1	S2	S3	Labour	Capital	HOUSEHOLD1		HOUSEHOLD2		Government	Total	Price
						CONH1	DEFH1	CONH2	DEFH2			
S1	7.66	42.11	17.04			17.97	0.54	5.99	0.38	10.14	101.82	1.204
S2	26.03	54.74	8.52			46.38	1.11	67.51	0.84	51.72	256.85	0.945
S3	14.55	36.21	4.26			4.09	0.99	14.37	1.10	7.92	83.48	0.927
Labour	32.69	46.58	19.12			16.77		89.85			205.00	0.718
Capital	57.99	77.39	34.62								170.00	0.552
Pollution ($\gamma = 10\%$)		25.685		Leisure price		0.646		0.646				
Unity damage	0.408			Utility index		100		100				

Table 6.3B. Social accounting matrix – economy with negative externalities.

Values	S1	S2	S3	Labour	Capital	HOUSEHOLD1		HOUSEHOLD2		Government	Total
						CONH1	DEFH1	CONH2	DEFH2		
S1	9.22	50.70	20.51			21.64	0.65	7.21	0.45	12.21	122.60
S2	24.60	51.73	8.05			43.83	1.05	63.80	0.79	48.88	242.72
S3	13.48	33.57	3.95			3.79	0.92	13.32	1.02	7.35	77.39
Labour	23.47	33.44	13.73			10.83		58.04			139.51
Capital	32.01	42.72	19.11								93.84
HOUS1				60.28	28.15					20.00	108.43
HOUS2				79.23	65.69					30.00	174.92
Government											118.43
Social security contribution	4.69	6.68	2.74								14.12
Excise duty											
S1	0.46	2.54	1.03			1.08	0.03	0.36	0.02		5.52
S2	1.23	2.59	0.40			2.19	0.05	3.19	0.04		9.69
S3	0.67	1.68	0.20			0.19	0.05	0.67	0.05		3.50
Value added tax											
S1						4.32	0.13	1.44	0.09		5.98
S2						8.74	0.21	12.73	0.16		21.84
S3						0.76	0.18	2.66	0.20		3.80
Corporate tax	12.81	17.09	7.65								37.54
Personal Income Tax						7.76		8.69			16.45
Total*	122.64	242.73	77.37	139.51	93.84	105.13	3.27	172.09	2.84	118.43	

*Numbers may vary due to rounding.

6.3.2. *Policies considered*

Four environmental policies are considered. The first is so-called voluntary abatement where the polluter unilaterally abates all emissions. The second is the Pigouvian tax where an eco-tax is set equal to the value of the total marginal damages imposed by the polluter. Results are shown in Table 6.4, respectively, in columns B and C. In general, labour taxes, output taxes and other distorting taxes can be viewed as having an indirect Pigouvian component. Such taxes lead to a reduction in the output level of the polluter and to an increase in its output price. We make differential incidence tests on pre-existing taxes. Therefore, the third policy considers the substitution of an additional tax levied on labour inputs and, finally, the fourth considers the substitution of an additional excise duty on output. Of course, the Pigouvian tax will generate a revenue equal to that of each replaced tax. The welfare cost of the tax-switching is computed according to the MCF formula (eqn. 6.20).

Voluntary abatement case

In this simulation the polluter voluntarily decides to abate all emissions by installing along side the productive process abatement technologies and by purchasing additional primary and intermediate inputs. The polluting firm's price increases as the abatement costs are internalized. The increase in the (ex)polluting commodity price generates an additional inter-commodity margin distortion and a cleaner environment which consequently increase the 'green' utility index. However this solution is apparently voluntary: it seems to be inefficient if the marginal abatement cost exceeds the marginal damage cost. This is true when the abatement ratio is higher than the optimal (i.e. Pigouvian) one which should be determined with reference to the value of marginal damages inflicted on the victims. The total abatement cost is equal to 30.5.

Pigouvian tax case

In this simulation the difference between marginal social cost and marginal private cost for each additional unit of polluting output is taxed. In other words, the Pigouvian tax forces the polluter to internalize the external costs and to emit a socially optimal amount of pollution. The model endogenously computes the optimal tax rate paid by the polluter and the optimal abatement level, and by implication the optimal pollution, according to equations (6.15) and (6.17) respectively. The Pigouvian tax t (eqn. 6.15) should be set equal to 0.927, the unity abatement level x (eqn. 6.17) will be equal to 0.037 and the unity damage level d (eqn. 6.2) will be equal to 0.067.

The direct cost of the Pigouvian policy to the polluting firm is equal to 20.249, i.e. the sum of the total abatement cost (15.624) the Pigouvian tax revenue (4.625) levied on residual pollution, and it is lower than the voluntary policy cost which is equal to 30.511.

A cleaner environment increases both the 'green' utility index and the

Table 6.4. Market equilibrium experiments.

Experiment	Voluntary policy B	Pigouvian policy C	Labour policy D	Output policy E
Utility Index				
Consumer 1	103.802	104.586	91.290	95.240
Consumer 2	100.021	101.733	97.964	95.034
Net pollution	0	16.91	24.803	25.610
Consumption (quantity)				
Consumer 1				
C1	24.289	22.955	17.332	17.862
C2	43.359	44.267	40.392	42.548
C3	4.41	4.255	3.883	4.055
Consumer 2				
C1	8.128	7.643	6.143	5.956
C2	63.385	64.388	62.54	61.98
C3	15.57	14.946	14.517	14.266
Defensive expenses				
Consumer 1				
S1		0.511	0.518	0.534
S2		0.778	0.954	1.005
S3		0.757	0.928	0.97
Consumer 2				
S1		0.291	0.383	0.372
S2		0.486	0.772	0.766
S3		0.695	1.104	1.086
Production (quantity)				
S1	113.05	119.15	101.90	103.78
S2	244.87	268.39	248.03	256.11
S3	85.44	90.71	83.57	84.95
Labour demand				
S1	21.859	23.174	33.979	32.916
S2	45.172	46.556	37.433	46.656
S2: abatement	8.645	1.192		
S3	19.922	19.846	20.833	19.552
Capital demand				
S1	37.686	44.279	52.582	57.95
S2	72.937	83.311	84.522	76.923
S2: abatement	24.332	3.719		
S3	35.045	38.691	33.896	35.122
Leisure				
Consumer 1	17.515	16.741	16.051	16.282
Consumer 2	91.887	97.491	96.704	89.594
Endowments				
Labour	205	205	205	205
Capital	170	170	170	170

Continued.

Table 6.4. (Continued.)

Experiment	Voluntary policy B	Pigouvian policy C	Labour policy D	Output policy E
Income				
Consumer 1	102.596	104.896	90.474	98.773
Consumer 2	172.081	170.634	158.676	163.121
Government tax revenue	121.518	138.901	127.125	129.857
Eco-tax revenue		15.624	15.625	14.999
Government expenditure				
S1	13.623	16.146	12.148	11.929
S2	48.051	61.519	55.938	60.031
S3	8.497	10.28	9.348	9.303
Unity damage *d*		0.067	0.38	0.398
Total abatement cost	30.511	4.625		
Unity emission level	0.1	0.1	0.1	0.1
Unity abatement level	0.1	0.037		
Pigouvian tax		0.927		
Tax on labour input			0.67	
Tax on polluting output				0.073
Equivalent variation				
Consumer 1	5.46	4.45	−9.00	−5.18
Consumer 2	0.02	2.79	−2.85	−7.61
MCF		−0.354	1.362	1.119
TEV		−1.266	0.474	0.131
Prices				
Producer prices				
S1	0.937	0.983	1.133	1.195
S2	1.063	1.032	0.985	0.950
S3	0.903	0.928	0.885	0.921
W	0.735	0.808	0.623	0.699
R	0.582	0.577	0.55	0.542
Consumer prices				
S1	1.171	1.228	1.416	1.493
S2	1.328	1.290	1.230	1.269
S3	1.128	1.159	1.106	1.151
Leisure				
Consumer 1	0.662	0.727	0.561	0.629
Consumer 2	0.662	0.606	0.561	0.629
Numeraire	1	1	1	1

Government tax revenue. The Pigouvian superiority is also shown by the increase in total Government revenue from 121.518, in the case of voluntary abatement policy, to 138.901.

Differential incidence between Pigouvian tax and labour input tax
The efficiency effects of the Pigouvian tax are compared with those obtained by increasing the mandatory social security contributions levied on labour input which is demanded by the polluting firm. The additional labour tax rate (ssc) should be set as equal to 0.52 per unit of labour input in order to satisfy the government revenue invariability condition.

A concise measure of the differential incidence efficiency is given by the marginal cost of public fund MCF (eqn. 6.20), which is given by the ratio between the equivalent variation and the change in the Government revenue. In the Pigouvian simulation the equivalent variation expresses households' willingness to accept in order to avoid the introduction of the Pigouvian fiscal reform. Therefore, MCF should be interpreted as a marginal revenue instead of a marginal cost. In this simulation the equivalent variation expresses households' willingness to pay in order to avoid the introduction of the tax on labour use. MCF correctly represents a marginal cost of the fiscal reform.

Differential incidence between Pigouvian tax and output excise duty
In this case, we compare the efficiency effects of a Pigouvian tax to those arising from an excise duty imposed on the polluting activity. The additional tax duty should be set equal to 0.07 per unit of output in order to satisfy the government revenue invariability condition.

According to the MCF indices, households should prefer the output tax policy to the increase in the mandatory social security contributions because the additional tax levied on polluting output is less distortionary.

Sensitivity analysis
We have varied the model's 'guesstimated' elasticities in order to test the sensitivity of the results. The model's response seems to be robust as the relative advantages and disadvantages of each of the policies are not affected by the changed elasticities.

6.3.3. *Summary of numerical comparisons and conclusions*

The Pigouvian policy (column C, Table 6.4) is preferable for two main reasons: it is Pareto optimal and it guarantees the maximum tax revenues when the effects of the negative externality are explicitly considered.

In the reference economy with externality but without an abatement policy (Table 6.3) the pollution level is a maximum (25.59). The utility index of both households is set equal to 100. Although the Pigouvian policy results in a positive pollution level (16.89), it determines the maximum 'green' utility index of both households (104.586 and 101.733).

It should be stressed that all policies have significant and different effects in terms of welfare. In fact the two consumers suffer tax burdens in different proportions. Comparison can be made looking at the equivalent variation rows in Table 6.4.

With regard to the producer prices, the Pigouvian policy results in the lowest price increase. Due to the effects of the negative externality the producer–victim's (S1) price raises significantly more than the polluting firm's (S2) price because the polluter is not forced to abate emissions by the fiscal policy (see columns D and E). Moreover, the polluting firm's price increases more in the case of the voluntary policy (column B) than in the case of the Pigouvian policy because the total abatement cost results in more than that achieved by Pigouvian tax.

As far as the tax switching policy results are concerned (row MCF, columns C, D and E), the Pigouvian tax provides a double dividend, in the strong form, of improvements in the environmental quality and also a reduction in distortionary costs of the fiscal system. The net cost of a cleaner environment, measured both in terms of MCF and TEV, is negative. This means that cuts in distortionary pre-existing taxes levied both on labour and output of the polluting firm are sufficient to achieve revenue-neutrality and to generate benefits in excess of gross cost. This conclusion derives from a utility function for the household which models leisure as an alternative to work. In a realistic model of the economy, where we suppose agents respond to incentives, it is possible to generate revenue neutral income by the state by tax burden reallocation in favour of the environment.

One of the justifications claimed for not implementing a tax switching policy is that it would reduce Government tax revenues and hence increase the Government deficit. Policy makers assert that a permanent reduction in a distortionary tax, such as a personal income tax or a mandatory social security tax on labour, can not be compensated by an environmental tax because, in the long run, economic agents will conform their production or consumption patterns to the environmental tax.

We think these critiques do not value costs and benefits correctly. In fact, if we compare tax revenue in the presence of negative externality against tax revenue in the absence of negative externalities, we can argue that any environmental policy actually leads a reduction in the tax revenue in which specific circumstances are true. In fact, total tax revenue in absence of negative externalities is equal to 142.79 (Table 6.2B) while it becomes equal to 121.518 in case of voluntary abatement policy (Table 6.4, column B), to 138.901 in case of Pigouvian policy (Table 6.4, column C), to 127.125 in case of labour policy (Table 6.4, column D), and finally to 129.857 in case of output' polic (Table 6.4, column E).

Results also suggest that, from the alternative environmental policies considered, Pigouvian policy is able to provide the highest level of tax revenue. When the effects of a negative externality are explicitly considered in the production and utility functions, however, any environmental policy that reduces (neutralizes) the external costs provides, in our competitive framework, a substantial increase in the overall tax base, compared with the benchmark which is equal to 118.43 (Table 6.3B).

Of course, this general result needs to be validated in a dynamic approach in order to consider both dynamic substitution among inputs and technical progress. In this case, we should be able to model not only 'end-of-pipe' technologies but also integrated technologies based on process change explicitly developed to reduce environmental effects.

Notes

1. I thank Andrea Beltratti, Christian M. Dufournaud, Alberto Majocchi, Karl-Göran Mäler, Alberto Pench, Sherman Robinson and an anonimous referee for valuable comments. Usual caveats apply.
2. Two methodological approaches are widely used for the evaluation of external costs. The first one (the so-called top-down approach) uses highly aggregated data to estimate average damage costs generated by specific pollutants. In contrast, the second approach (the so-called bottom-up approach) uses technology specific emissions data and dose–response functions to calculate the physical and external cost impact of incremental emissions and damages. The bottom-up approach should be preferred because the top-down approach is not suited to the estimation of marginal external costs, for which a detailed understanding is necessary to evaluate Pigouvian taxes.
3. Four distinct measures, which have the potential to reduce pollutant emissions from road vehicles, are considered: technical measures, by improving vehicle technology; fuel quality measures, by enhancing the quality of fuel; emission standard measures, by setting inspection and maintenance programmes; non-technical measures, by promoting the most economically and environmentally efficient use of road space. Non-technical measures include improved traffic management (i.e. road pricing, traffic restrictions, speed regulations), urban transport enhancements, and instruments aimed at influencing vehicle ownership and use (i.e. fuel tax, vehicle purchase tax, and scrappage subsidy). The European Union proposal of an energy-carbon tax was the first concrete policy proposal for the direct use of an economic instrument in environmental policy.
4. In the economic literature the 'double dividend' notion, introduced by Pearce (1991, p. 940), motivated a large debate. In the American debate a reduction in distortionary costs of the fiscal system in addition to improved environmental quality is termed a double dividend. On this issue see Goulder (1995). In this European debate an increase in employment due to an environmental taxation is generally regarded as an important extra divident in addition to improved environmental quality. On this version of the double dividend, see Majocchi (1996).
5. Quoting Goulder (1995, p.158), "... Much of the controversy about the second dividend is in terms of whether environmental taxes can be introduced in a way that is costless. Policy-makers who are interested in green tax swaps are often frustrated by uncertainties as to the values of the environmental benefits that would result from such swaps. Under these conditions the no-cost idea is especially appealing. If the revenue-neutral environmental tax policies are costless, the burden of the proof facing the policy-maker is much reduced: it is sufficed to know the sign of the environmental benefits – to know that they are positive. If costs are zero (or negatives), this guarantees positive net benefits."
6. The reader is referred to Adams and Crocker (1991) for an introduction to this issue.
7. If the victim's output is reduced by q ($q = 2, , \ldots, m$) polluters, the exponent between brackets will be reset according to specific dose–response relationships, i.e.:

$$d(\ldots, Q_q, \ldots, \bar{\gamma}_q, \ldots, x_q, \ldots) = z\left(1 - \prod_q e^{-\phi_q \cdot Q_q(\bar{\gamma}_q - x_q)^{\theta_q}}\right) \quad 0 \leq d \leq z$$

In a dynamic approach pollution generated at period t should be added to the existing pollution stock in order to determine the total effect that damages the victim's output.

8. Nordhaus' DICE model shows a similar damage function in which the relationship between global temperature increase T and global output loss D is given by:

$$D_t = \theta_1 \cdot T_{I,t}^{\theta_2} \cdot Y_t$$

where q_1 is a parameter representing the scale of damage, q_2 represents the nonlinearity in the damage function and Y_t is global income (Nordhaus, 1994).

9. Samuelson (1969) has included private and public goods in the utility function and also the possibility of lump-sum transfers among households, determined optimally via a social welfare function within a 'pseudo' general equilibrium framework. Samuelson's equilibrium means that the marginal utility of each household's revenue (evaluated in terms of the burden that each household represents in the social welfare function) is identical. If it is possible to increase one household's revenue by means of lump-sum transfers in such a way as to equalise interpersonal equity among households it is possible to determine Pareto's 'bliss point'. See Pireddu (1996) for numerical simulation in the case of provision of public goods.
10. A neo-classical production function with emissions among traditional inputs was suggested by Brock (1977). He claims that any abatement technology uses inputs that must be taken away from the primary production. If it is possible to discharge pollutants with no costs the polluter, according to the neo-classical theory of substitution, could limit output by reducing costly inputs and raise emissions which have no market price. Therefore Brock's production function does not avoid the possibility that the abatement/output ratio can reach infinity (for more details, see Beltratti, 1993).

References

Adams, D.M. and T.D. Crocker, 1991, Material damages, in J.B. Braden and C.D. Kolstad, eds., Measuring the Demand for Environmental Quality, Amsterdam: North-Holland.

Aprile, G., 1984, Les depenses publiques en Suisse, Genève: Droz.

Ballard, C.L. and S.G. Medema, 1993, The marginal efficiency effects of taxes and subsidies in the presence of externalities, Journal of Public Economics, 52, 199–216.

Baumol, W.J., 1974, Environmental protection and income distribution, in H. M. Hochman and G.E. Peterson, eds., Redistribution through Public Choice, New York: Columbia University Press.

Beltratti, A., 1993, La contabilita' ambientale: aspetti teorici e implicazioni per la politica economica ambientale, in I. Musu and D. Siniscalco, eds., Ambiente e contabilita' nazionale, Bologna: Il Mulino.

Brock, W., 1977, A polluted golden age, in V.L. Smith, ed., Economics of Natural and Environmental Resources, New York: Gordon and Breach.

Burrows, P., 1986, Nonconvexity induced by external costs on production: theoretical curio or policy dilemma?, Journal of Environmental Economics and Management, 13, 101–128.

Conrad, J.M. and C.W. Clark, 1987, Natural Resource Economics. Notes and Problems, Cambridge: Cambridge University Press.

European Commission, DG III, 1995, A cost-effectiveness study of the various measures that are likely to reduce pollutant emissions from road vehicles for the year 2010. Final Report.

European Commission, DG XII, 1994, Externalities of Fuel Cycles. 'ExternE' Project. Summary Report.

Goulder, L.H., 1994, Energy taxes: traditional efficiency effects and environmental implications, in J. Poterba, ed., Tax Policy and the Economy, Cambridge: MIT Press.

Goulder, L.H., 1995, Environmental taxation and the double dividend: a reader's guide,

International Tax and Public Finance, Special Issue on Public Finance and the Environment, Dordrecht: Kluwer.
Majocchi, A., 1996, Green fiscal reform and employment: a survey, Environmental and Resource Economics, 8, 375–397.
Nordhaus W.D., 1994, Managing the Global Commons. The Economics of Climate Change, Cambridge: MIT Press.
Pavan, M., 1995, Environmental damage valuation: an Italian case-study, paper presented at Terza Riunione Scientifica degli Economisti Ambientali Italiani, Pavia: Universita' degli Studi.
Pearce, D.W., 1991, The role of carbon taxes in adjusting to global warming, Economic Journal, 101, 938–948.
Pireddu, G., 1996, Pollution, externalities and optimal provision of public goods: a computable general equilibrium approach, in A. Fossati, ed., Economic Modelling Under the Applied General Equilibrium Approach, Aldershot: Avebury.
Rossier, E., 1976, Sur le comportement de l'etat dans le cadre de la theorie des choix, in L. Solari and J.-N. du Pasquier, eds., Private and Enlarged Consumption, Amsterdam: North-Holland.
Samuelson, P., 1969, Pure theory of public expenditure and taxation, in J. Margolis and H. Guitton, eds., Public Economics, An Analysis of Public Production and Consumption and their Relations to the Private Sectors, London: MacMillan.
Shoven, J.B. and J. Whalley, 1992, Applying General Equilibrium, Cambridge: Cambridge University Press.

Appendix 1. Producer's damage cost function

In order to satisfy the convexity assumption of the conventional theory (see Burrows, 1986) the first and second-order conditions of the external cost function, given in equation (6.3) of the text, should be:

$$\frac{\partial EC_{Q_{S1}}}{\partial x} < 0 \quad \text{and} \quad \frac{\partial^2 EC_{Q_{S1}}}{\partial x^2} > 0.$$

To show that the marginal external cost function has the proper sign it is sufficient to derive the damage function:

$$\left.\frac{\partial d}{\partial x}\right|_{0 \le x < \bar{\gamma}} = -z\phi\theta Q_{S2}(\bar{\gamma} - x)^{\theta-1} \cdot e^{-\phi Q_{S2}(\bar{\gamma}-x)^{\theta}} < 0.$$

This first partial derivative is equal to zero when $x = \bar{\gamma}$, the upper limit of the abatement variable, and $\theta \neq 1$.

Figure 6.A1 shows the first partial derivative function for the values of the parameters selected in the text. Particularly $\bar{\gamma}$ was set equal to 0.1 and θ to 2.

The second derivative of the damage function is

$$\left.\frac{\partial^2 d}{\partial x^2}\right|_{0 \le x < \bar{\gamma}} = z\phi Q_{S2}\theta \cdot e^{-\phi Q_{S2}(\bar{\gamma}-x)^{\theta}} \cdot (\bar{\gamma} - x)^{\theta-2}$$
$$\times [(\theta - 1) - \phi Q_{S2}\theta(\bar{\gamma} - x)^{\theta}] \; ?> 0.$$

The necessary condition for a positive sign is $\theta > 1$ as

$$\frac{\theta - 1}{\phi Q_{S2}\theta} > (\bar{\gamma} - x)^{\theta}$$

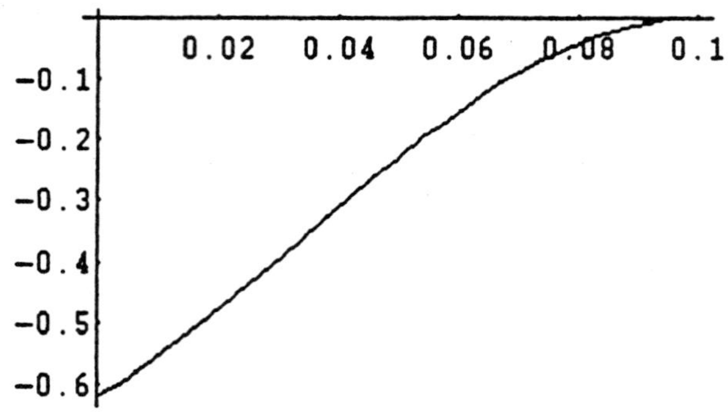

Figure 6A.1.

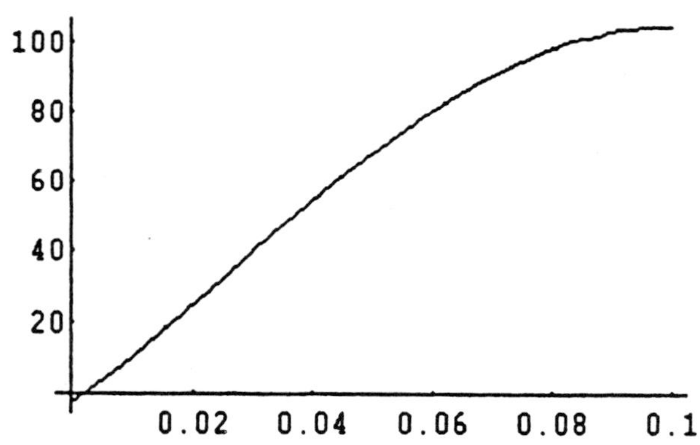

Figure 6A.2.

The condition is not sufficient because the sign of the second derivative can become negative for abatement values very close to zero and for particular values of the polluting output Q_{S2} and of the parameters γ and ϕ. Moreover, when $x = \bar{y}$ and $\theta = 2$ the second partial derivative is undetermined. Figure 6.A2 plots the second derivative function for the values of the parameters selected in the text.

If the damage function becomes concave it fails to meet the requirements of convexity of the conventional theory. But in our simulations the nonconvexity seems to be 'irrelevant' because it is not an obstacle to an optimal pollution control policy.

Appendix 2. Household's defensive expenditure

A household's defensive expenditure, given in equation (6.10) of the text, is

$$EC_h = PAGGD_h DEFC_h$$

where $PAGGD_h$ is the implicit price of defensive expenditure, and

$$DEFC_h = (1 - \beta_{0h} - \beta_{1h})(\gamma - x)^{[\tau(\sigma-1)+1]/\tau} \cdot Y_h/(PAGGD_h^{\sigma} \cdot \Delta_h)$$

$$\Delta_h = PAGGC_h^{(1-\sigma)}\beta_{0h} + P_{lh}^{(1-\sigma)}\beta_{1h} + PAGGD_h^{(1-\sigma)}(1 - \beta_{0h} - \beta_{1h})(\gamma - x)^{[\tau(\sigma-1)+1]/\tau}$$

To show the first and second-order partial derivative it is sufficient to derive the defensive expenditure demand, DEFC. The expected signs of the first and second-order conditions are both negative. Let us simplify the notation in the following manner:

$$a = (1 - \beta_{0h} - \beta_{1h}) > 0$$

$$b = \frac{Y_h}{PAGGD_h^{\sigma}} > 0$$

$$c = PAGGC_h^{(1-\sigma)}\beta_{0h} + P_{lh}^{(1-\sigma)}\beta_{1h} > 0$$

$$d = PAGGD_h^{(1-\sigma)}(1 - \beta_{0h} - \beta_{1h}) > 0$$

$$n = \frac{\tau(\sigma - 1) + 1}{\tau} > 0; \quad \tau > 0 \quad \text{and} \quad 0 < \sigma < 1$$

While a, b, c and d are greater than zero by definition, n is positive if τ is in the following range: $0 < \tau \leq \tau_{max} = 1/1 - \sigma$. This condition states the range of the parameter τ with respect to the elasticity of substitution parameter σ. Remember that τ and σ are both 'guesstimated' parameters in our CGE model.

The first partial derivative is

$$\left.\frac{\partial DEFC_h}{\partial x}\right|_{0 \leq x < \bar{\gamma}} = -\frac{abn(\bar{\gamma} - x)^{n-1}}{c + d(\bar{\gamma} - x)^n} + \frac{abnd(\bar{\gamma} - x)^{2n-1}}{(c + d(\bar{\gamma} - x)^n)^2} \;?< 0$$

The sign is negative as

$$-1 < \frac{d(\bar{\gamma} - x)^n}{c + d(\bar{\gamma} - x)^n}$$

Figure 6.A3 shows the first derivative function for the values of the parameters selected in the text.

The second partial derivative is

$$\left.\frac{\partial^2 DEFC_h}{\partial x^2}\right|_{0 \leq x < \bar{\gamma}} = \frac{abn(n-1)(\bar{\gamma} - x)^{n-2}}{c + d(\bar{\gamma} - x)^n} - \frac{abdn^2(\bar{\gamma} - x)^{2n-2}}{(c + d(\bar{\gamma} - x)^n)^2} - \frac{abdn(2n-1)(\bar{\gamma} - x)^{2n-2}}{(c + d(\bar{\gamma} - x)^n)^2} + \frac{2abd^2n^2(\bar{\gamma} - x)^{3n-2}}{(c + d(\bar{\gamma} - x)^n)^3} \;?< 0$$

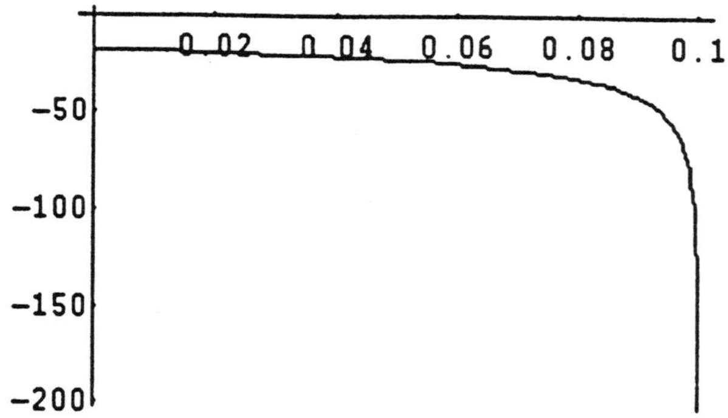

Figure 6A.3.

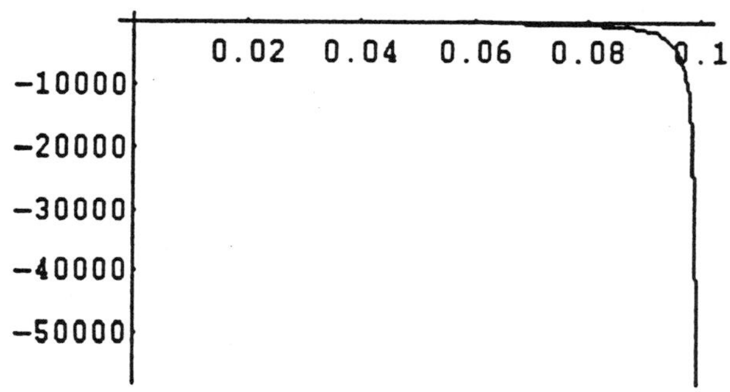

Figure 6A.4.

The expected sign of the second derivative is negative. As this condition is not easy to handle, we show (Figure 6.A4) that the values of the selected parameters of our model guarantee the expected sign.

Box 1. Equations of the basic CGE model with negative externalities.

Equation	Variable	Explanation
$Q_j = c_j L_j^{\alpha_{j1}} K_j^{1-\alpha_j}$	Q_j	Sectoral Cobb–Douglas production functions ($j =$ S2, S3). Q is output, L is labour, K is capital, c and α are parameters.
$Q_{S1} = c(1-d)L^{\alpha}K^{1-\alpha}$	Q_n	Cobb–Douglas production function of the victim (S1). d is the damage (dose–response) function.

$d = z(1 - e^{-\phi Q_{S2} \cdot (\bar{y} - x)^{\theta}})$	d	Damage function (see section 2.2.2)
$NP_i = P_i - \sum_j A_{ji} \cdot P_j$	NP_i	Value added or net price. P is the sectoral price and A_{ji} are input–output coefficients. $i = j = S1, S2, S3$
$w(1 + ssc_i) = NP_i \cdot \alpha_i \cdot \frac{Q_i}{L_i}$	L_i	Sectoral labour demand equations. w is the wage rate $i = S1, S2, S3$
$r(1 + corp_i) = NP_i \cdot (1 - \alpha_i) \cdot \frac{Q_i}{K_i}$	K_i	Sectoral capital demand equations. r is the capital rental rate. $i = S1, S2, S3$
$CON_h = \beta_{0h} Y_h / (PAGGC_h^{\sigma} \cdot \Delta_h)$	CON_h	CON is the composite commodity of each household's normal consumption. $h = 1, 2$ (see section 2.2.3)
$L_h = \beta_{1h} Y_h / (P_{lh}^{\sigma} \cdot \Delta_h)$	L_h	Household's leisure demand. $h = 1, 2$ (see section 2.2.3)
$DEFC_h = \frac{(1 - \beta_{0h} - \beta_{1h})(\gamma - x)^{[\tau(\sigma - 1) + 1]/\tau} \cdot Y_h}{(PAGGD_h^{\sigma} \cdot \Delta_h)}$	$DEFC_h$	DEFC is the composite commodity of each household's defensive consumption. $h = 1, 2$ (see section 2.2.3).
$C_{ih} = l_{ih} \cdot PAGGC_h \cdot CON_h / MP_i$	C_{ih}	Normal goods conditional demands. $i = S1, S2, S3$; $h = 1, 2$
$CD_{ih} = e_{ih} \cdot PAGGD_h \cdot DEFC_h / MP_i$	CD_{ih}	Defensive goods conditional demands. $i = S1, S2, S3$; $h = 1, 2$
$MP_i = P_i(1 + tax_i)(1 + vat_i)$	MP_i	Consumer prices. $i = S1, S2, S3$
$PAGGC_h = \prod_i \left(\frac{MP_i}{\lambda_{ih}}\right)^{\lambda_{ih}}$	$PAGGC_h$	Household's implicit price of the normal composite goods. $i = S1, S2, S3$; $h = 1, 2$
$PAGGD_h = \prod_i \left(\frac{MP_i}{\varepsilon_{ih}}\right)^{\varepsilon_{ih}}$	$PAGGD_h$	Household's implicit price of the defensive composite goods. $i = S1, S2, S3$; $h = 1, 2$
$P_{lh} = w(1 - PIT_h)$	P_{lh}	The price of leisure is set equal to the wage rate net of household's highest marginal income tax rate. $h = 1, 2$

$Y_h = w \cdot \sum_i L_i \cdot EL_h \cdot r \cdot \sum_i K_i \cdot EK_h + TRASF_h - IT_h$	Y_h	Total household's income. EL is the endowment of labour and EK is the endowment share of capital share to each household. TRASF is the exogenous net transfers from the government and IT is the total personal income tax; $h = 1, 2$.
$IT_h = BLOCK1 \cdot 0.1 + BLOCK2 \cdot 0.25$	IT_h	Total personal income tax according to two progressive tax rates. $h = 1, 2$
$GOVREV = \sum_h IT_h + \sum_i corp_i \cdot r \cdot K_i + \sum_i ssc_i \cdot w \cdot L_i + tax_i \cdot \sum_j P_i \cdot A_{ji} \cdot Q_j + tax_i \cdot \sum_h P_i \cdot (C_{ih} + CD_{ih}) + vat_i \cdot \sum_j P_i(tax_i) \cdot A_{ji} \cdot Q_j + vat_i \cdot \sum_h P_i(tax_i) \cdot (C_{ih} + CD_{ih})$	$GOVREV$	Total government tax revenue. $i =$ S1, S2, S3; $h = 1, 2$
$GOVEXP_i = g_i \cdot \dfrac{\left(GOVREV - \sum_h TRASF_h\right)}{P_i}$	$GOVEXP_i$	Sectoral government expenditure of Cobb–Douglas type. $i =$ S1, S2, S3
$C_i + CD_i = Q_i - \sum_j A_{ji} \cdot Q_j - GOVEXP_i$	P_i	Product market balance equation. $i = j =$ S1, S2, S3
$\sum_i L_i + \sum_h L_h = \bar{L}$	w	Labour market balance equation. $\bar{L}$ is the exogenous supply of labour; $i =$ S1, S2, S3; $h = 1, 2$
$\sum_i K_i = \bar{K}$	r	Capital market balance equation. $\bar{K}$ is the exogenous supply of capital. $i =$ S1, S2, S3
$\bar{P} = \prod_i P_i^{\delta_i}$		Numeraire price index. $\bar{P} = 1$ and $\Sigma_i \delta_i = 1$. $i =$ S1, S2, S3

There are 54 equations, 53 endogenous variables and 2 exogenous variables. Therefore, a clearing condition must be dropped because of Walras' law.

Box 2. Variables of the basic CGE model with negative externalities.

Endogenous variables	Number	Explanation
		$i=j=$ S1, S2, S3
		$h=1,2$
Q_j	3	Sectoral output
d	1	Unity damage
NP_j	3	Sectoral net price
L_j	3	Sectoral labour demand
K_j	3	Sectoral capital demand
CON_h	2	Household's standard' composite good demand
$DEFC_h$	2	Household's defensive composite good demand
L_h	2	Household's leisure demand
C_{ih}	6	Household's standard commodity demand
CD_{ih}	6	Household's defensive commodity demand
MP_i	3	Consumer prices
$PAGGC_h$	2	Household's implicit price of the standard composite good
$PAGGD_h$	2	Household's implicit price of the defensive composite good
P_{lh}	2	Leisure price
Y_h	2	Household's expanded income
IT_h	2	Household's personal income tax
$GOVREV$	1	Total tax revenue
$GOVEXP_i$	3	Sectoral government expenditure
P_i	3	Sectoral producer price
w	1	Wage rate
r	1	Capital rental rate

CHAPTER 7

Economic Incentive Policies under Uncertainty: The Case of Vehicle Emission Fees[1]

Winston Harrington, Virginia McConnell and Anna Alberini

7.1. Introduction

Although economists have been recommending more reliance on economic incentive policies for reducing pollution for years, it is only recently that these policies have gained serious attention by policy makers. Economic incentive policies currently in use include marketable permits for electric utility emissions, some deposit refund recycling systems and the low emission vehicle programme in California. Cost savings from the use of incentive-based policies have been well-documented, but there may be many obstacles in practice to the implementation of such fees. In this paper, we examine a proposed new economic incentive policy, emissions fees on motor vehicles, which can be considered as a substitute for the current regulatory policy of vehicle inspection and maintenance. Although emissions fees for vehicles have been suggested in the literature (see Harrington et al., 1995; Kessler and Schroeer, 1993; White, 1982), some practical issues may prevent fees from having large advantages over more traditional regulations. Using a simulation model based on empirical evidence about vehicle emissions and repair effectiveness, we compare the net welfare impacts of emissions fees to a strict regulatory policy under which all polluting vehicles must be fixed.

The difference between economic incentive policies and more rigid regulatory policies will depend on how well the incentive policies are targeted and on their ability to influence behaviour. Fullerton (1995) has shown that most current pollution taxes are lump sum taxes designed to raise revenues, and are not well targeted to providing incentives to reduce emissions. The vehicle emissions tax under consideration here is a tax rate on grams of pollutant per mile, and would appear to be well targeted. A vehicle owner can respond to an emissions fee by repairing the vehicle, by driving it less, or by scrapping it. Our focus here is on the decision to repair a vehicle.

R. Roson and K.A. Small (eds.), Environment and Transport in Economic Modelling, 152–182.

The extent to which an emissions tax will affect consumer behaviour to reduce emissions depends on a number of factors including the underlying economic response (the true elasticities), how the fee is implemented (including timing and enforcement), and the extent of uncertainty associated with various aspects of its use. In this paper we include stochastic components for vehicle emissions, emissions measurement, and ability to predict the effect of repair on emissions. The distributions are all based on empirical data. We model the owner's decision about whether and how to repair a vehicle and use this in a simulation model of inspection and repair to evaluate alternative policies for reducing emissions.

An emissions fee has the potential to substantially increase economic welfare over current regulatory policy for reducing vehicle emissions, inspection and maintenance programmes (I/M). Under current I/M policy motorists must repair a vehicle if it does not pass the inspection, regardless of the benefits of doing so. With an emissions fee the motorist has a choice about whether to repair the vehicle and costly repairs can be avoided if they are unlikely to produce significant emissions reductions. However, uncertainty about emission levels and predictions about the effectiveness of repair can affect the welfare advantage of fees over more the more rigid I/M approach. For example, a motorist could be no better off if he consents to a repair that badly fails to achieve its anticipated emission reductions. Even when it has a welfare advantage, a fee policy can impose very high costs on individual motorists. Measures that mitigate these impacts can, unfortunately, compromise the efficiency objectives.[2]

Using the simulation model, we are also able to examine other aspects of the fee or regulatory policies. One major concern of policy makers is the distributional impact of these policies. Low income households are more likely to drive older, more polluting vehicles which will require more expensive repairs or larger fees or both (Walls and Hanson, 1995). As a result, most regulatory I/M programmes have upper limits on the amount that motorists can be required to pay for vehicle repair. These limits prevent individual motorists from bearing a particularly heavy burden, but they also can limit the effectiveness of I/M policies by allowing motorists to avoid cost-effective emission repairs.

There are a number of ways in which emission fees can be designed to reduce these distributional burdens in potentially cost-effective ways. One policy examined here is to use a two-part fee where emissions up to some level can have a low or zero fee rate, and emissions over the limit have a higher fee.[3] Under certainty, this fee design is likely to be less efficient than a single fee, but under uncertainty, the results are not as clear. An emissions fee can also be designed to cap the individual's maximum obligation, similar to caps set in most I/M programmes. Collected revenues can then be used to subsidize the cost of repairs beyond this cost limit.[4] The simulation results presented below suggest that these cost limits can be set at surprisingly low levels without exhausting the emission fee revenues.

The paper first gives some background on regulation of vehicle emissions in the USA. We then describe the simulation model and the underlying empirical relationships and assumptions. We then present results of the simulation model comparing emissions fees to command and control policies, focusing on the impact of uncertainty and on the distributional impact of the policies. Finally, we offer some conclusions and directions for future research.

7.2. Background on regulation of vehicle emissions in the United States

Despite intense regulatory efforts to reduce vehicle emissions over the past 20 years, emissions from cars and trucks continue to be major source of air pollution problems in urban areas of the USA (National Research Council, 1991). Until recently, the principal approach to reducing vehicle emissions had been stringent new car emission standards, enforced against the manufacturer. Although the new car emission reductions have been substantial, emissions from vehicles on the road continue to be a problem, with relatively small numbers of cars with broken or deteriorated emissions equipment accounting for the majority of the emissions.

Recognition of this divergence between new vehicle emission certification and actual in-use emissions caused Congress to require vehicle inspection and maintenance (I/M) programmes in the more polluted urban areas as part of the 1977 Clean Air Amendments. At the time, evidently, legislators envisioned emission inspection to be a straightforward extension of vehicle safety inspection programmes in operation in all the states and previously mandated by federal legislation. Many states, in fact, responded by tacking the I/M test onto the existing safety inspection.

States were given wide latitude for designing their own inspection programmes and a variety of programmes emerged. However, by the late 1980s it had become clear that many of the initial state programmes, on which the EPA had placed such high expectations, were not effectively reducing vehicle emissions and urban areas in many states were still not in compliance with national clean air standards. Many states did not appear to be enforcing I/M requirements. For example, most programmes had an upper limit on the amount motorists were required to pay for vehicle repair (some states set limits as low as $15 per vehicle). If, after expenditure up to the limit, the vehicle still could not pass the emission test, the owner was granted a waiver that permitted the vehicle to be operated even though its emissions exceed the limit. Many states also allowed certain model year vehicles to be exempt from the inspection and repair process, including, for example, vehicles more than 12 years old. These older vehicles often have the highest emissions rates.

Criticism of I/M programmes grew, as evidence accumulated about the inadequacy of existing programmes. One study of California's 'smog check' programme, a decentralized I/M system consisting of some thousands of privately owned gas stations, repair shops and dealerships certified by the State to conduct vehicle inspections, found evidence that cars appeared to be as dirty after smog check as they were before (Lawson, 1993). Apparently, the uncertainty and incomplete reproducibility of the emission tests, the inability of some emission tests to replicate actual driving conditions, the variability in true vehicle emissions, and the frequent ineffectiveness of emission repairs have provided motorists and mechanics ample opportunity to 'game' the system.

Gaming has taken a number of forms. There is evidence that motorists have failed to register their vehicles at all, or have registered them at false addresses in jurisdictions without I/M programmes. To avoid repair, motorists have made changes to their vehicles before the emissions test, and then undone those changes afterward. At the inspection site, mechanics can falsify test results or 'clean-pipe' vehicles, where the emission tests of a clean car are reported in place of the true results of a dirty car. Some mechanics argue that motorists also have opportunities for making inexpensive repairs that produce clean emission tests but which do not last. If a motorist fails a test, for example, a change of spark plugs may enable the vehicle to pass the retest, but the condition that caused the plugs to foul may be uncorrected and will soon lead again to high emissions. Finally, inasmuch as emission tests results vary from one test to the next, motorists also have the opportunity to pass the test simply by retaking the test. In general, these problems were exacerbated by fairly low limits on the amount motorists could be required to spend.

When the Clean Air Act was amended in 1990, Congress tightened the requirements for I/M in the most polluted urban areas, requiring enhanced I/M, which required that states do not set minimum repair limits at levels less than $450 per car, promoted centralized testing system with a more rigorous and extensive test procedure, and established strict restrictions on which cars have to be tested.[5] The EPA believed that the new enhanced I/M would improve some of the technical difficulties with the earlier tests and help to prevent at least some of the types of fraud.

These new regulations were to have been implemented by 1994, but there has been such opposition to them in so many states that the EPA has backed away from the original requirements that programmes should be centralized and use specific test technology. Federal regulators are considering more flexibility in how they allow states to comply with clean air statutes. Economic incentive policies like an emission tax will have considerable appeal if they can be implemented at lower cost and with greater effectiveness compared with more rigid regulatory policies. In addition, many other countries around the world not as locked into existing regulatory policies are considering economic incentives, including some types of emissions fees (Eskeland and Devarajan, 1996; S. Smith, 1995).

7.3. The simulation model

To examine the welfare implications of various emissions fee and regulatory policies for reducing vehicle emissions, we have developed a simulation model that captures aspects of both the stochastic and behavioural elements of emissions measurement and repair. The simulation uses a 1000-vehicle fleet, in which vehicle emissions and annual mileage are randomly drawn from age-specific distributions and distributional parameters reflecting the California vehicle fleet in 1991. To identify a vehicle's emissions, we assume the vehicle receives an emissions test, the results of which measure the 'true' emission rate with error. Based on test results and the type of command and control or fee policy in use, the vehicle may or may not be repaired. Vehicles which are repaired are then retested to determine the post-repair emission rate.

Stochastic elements of the model include the initial true emission level, the emission measurement error, the effectiveness of vehicle repairs, and the ability of mechanics to predict repair effectiveness. The simulation relies on data taken from a number of empirical studies: emissions data on HC, CO and NO_x is from I/M readings taken by EPA at a test lane in Indiana; test accuracy is characterized by data from the California I/M Review Committee undercover car study (California I/M Review Committee, 1993); finally the data for repair effectiveness derive from two separate studies: the same California undercover car study and the other the Sun Oil Company scrap-repair study (Cebula, 1994). These data are used to estimate statistical models of emission measurement and vehicle repair that generate the parameters of the probability distributions used in the simulation.

There are both policy parameters which can be changed to reflect different types of fees or regulatory policies, and technical parameters which can be changed to reflect the degree of uncertainty in the model. Two technical parameters of special interest are those that describe the precision of the emissions measurement and the ability to predict repair effectiveness. One criticism of emissions fees is that if there is uncertainty about a vehicle's emissions rates, then a fee cannot be well targeted. However, uncertainty can also affect the performance of the regulatory programme as well. We use the simulation model to vary the technical parameters and compare the results of the two policies. The ability of mechanics to predict repair effectiveness is particularly important for emissions fees, because this prediction forms the basis of the motorist's decision about whether to repair the vehicle. The model also allows for parameter changes that simulate improvement in mechanics' predictive ability.

We will also use the model to compare other aspects of emissions fees and the more rigid regulatory policy which we will call command and control (C&C) policies. The C&C policy we examine will be an I/M programme that tests all vehicles, and requires repair on any that fail the test cutpoint.[6] Variants of this policy include a limit on the amount motorists have to pay for repairs, such as the $450 minimum limit specified by the 1990 Clean Air Act (see section 2). The emissions fee policy would have a similar structure, in that vehicle emissions would have to be identified. The difference would be that owners with vehicles whose emissions exceed some limit, called the fee baseline, would be allowed to pay a fee based on the emission test results instead of making required repairs. On retest, the vehicle would still be liable for emission fees if the test results exceeded the baseline. The repair decision, then, would be based on a comparison of the fee before repair with the sum of the repair cost and the expected fee, if any, after repair.

The model is capable of including a number of responses to fees or I/M policies, including repair, vehicle scrappage, or changes in driving behaviour. In this paper we focus on the repair response.[7] The fee policies to be analysed are all emission rate fees; motorists cannot reduce their fees by driving less.[8] A limitation of the model is that it is static, and therefore does not incorporate the dynamics of vehicle emission deterioration and fleet turnover, preventing consideration of such issues as the durability of repair and motorists' maintenance incentives.[9] For calculating the benefits of alternative policies we assume that the duration of repairs is one year.

7.3.1. *The model: emissions distributions, measurement and repair effectiveness functions*

Initial emissions distribution

The simulation is applied to a 1000-vehicle fleet with emissions randomly drawn from an emissions distribution resembling the fleet of vehicles in Southern California in 1991 (see Table 7.A1 of the Appendix for a summary of the emissions data).[10] The underlying data for this distribution were obtained from remote sensing readings of over 90 000 vehicles from California roads over a 6-month period (Bishop and Stedman, 1994). This dataset had readings only for HC and CO emissions, so NO_x readings had to be inferred from a different data source from the EPA.[11] The resulting joint emissions distribution represents the true emissions of the vehicles.

Accuracy of emissions test

All vehicles must have an emissions test to identify emission rates and we assume here that vehicles are tested with the current test equipment in place in California, the BAR-90 test, a two-speed test that includes, in addition to a simple idle test, a test of emissions at an engine speed of 2500 rpm. Given that the test cannot perfectly measure the true emissions of the vehicle, we need to know how well it does. For each pollutant, we estimate the relationship between the BAR-90 test and a more complete emission test, the FTP which we take to be a measure of the vehicle's true emissions,[12] using data from a study of 681 cars in California (California I/M Review Committee, 1993).

Table 7.1. Repair effectiveness in non-EPA empirical studies.

	N	Average cost	E_0 (g/mile)	E_1 (g/mile)
California I/M Review Committee (1993)	681*	\$89.55		
HC			3.11	2.32
CO			37.21	29.4
NO_x			1.67	1.53
Sun Oil Company (Cebula, 1994)	155	\$338.55		
HC			4.83	1.55
CO			69.18	17.00
NO_x			2.90	2.64
Total Petroleum (Lodder and Livo, 1994)	103	\$390.21		
HC			3.66	2.48
CO			45.64	33.38
NO_x			not given	not given

The general form is:

$$FTP_i = \mathbf{Z}\alpha + \varepsilon \qquad (7.1)$$

where FTP_i refers to the FTP test emissions for i = HC, CO and NO_x, in g/mile, and $\mathbf{Z}$ refers to a vector of measurements from the BAR-90 test.[13] These equations provide a measure of the accuracy of the current test for identifying a vehicle's true emissions. Estimation results for equations (7.1) for HC, CO and NO_x are summarized in Table 7.A2 of the Appendix. We use the results in the simulations described below.

Repair effectiveness

The simulation procedure will also require estimation of the effectiveness of repair. There is currently a great deal of controversy about both the cost and effectiveness of emissions system repairs. The EPA's original estimates of repair effectiveness were very optimistic, assuming emissions would be reduced 45–90% after repair. However, recent studies show vehicle emission repairs to be much less successful and more expensive than the EPA projections assumed. One of those studies was the undercover car study described above (California I/M Review Committee, 1993). Table 7.1 shows that the average cost of repair was low, less than \$90 per vehicle on average, probably a result of waiver limits that restricted the amount spent on each vehicle. The average emission reduction was also low, 25% for HC, 21% for CO and 8% for NO_x. Emission reductions were also variable, with nearly half the cars showing higher emissions after repair than before, and were not at all related to cost. The poor results may also be attributable in part to the way the programme was administered by the BAR in California, where, according to some mechanics, it is easier to get in trouble with the state by presenting the motorist with a large repair bill than it is by failing to produce a genuine emission reduction.[14]

In two other recent repair studies the repairs were more effective, but far more expensive than assumed by the EPA. In 1993 the Sun Oil Company (Cebula, 1994) used remote sensing to identify gross-emitting vehicles owned by their employees as they left company parking lots in Philadelphia. Sun offered to repair these vehicles at its expense, and the company spent up to \$450 on each vehicle. As shown in Table 7.1, after an average expenditure of \$338, emissions of HC, CO and NO_x were reduced by 68%, 75% and 9%, respectively, showing significant improvement over the California results.[15] A second study was performed by Total Petroleum in Denver (Lodder and Livo, 1994), in which gross-emitting vehicles were identified and repaired at Total Petroleum's own expense. As shown, emissions of HC and CO were reduced by about a third after average expenditure of nearly \$400.

In the simulations below we need to able to predict by how much a given repair will reduce vehicle emission rates. If mechanics had this information *ex ante*, they could use it to select the combination of repairs that would bring the vehicle into compliance with the emission standards at least cost. Repair predictability would be particularly useful in an emission fee programme, since the mechanic can decide on the package of repairs that maximizes the expected net benefit of repair to the motorist – that is, the expected reduction in emission fees paid less the cost of repair. This choice can include the possibility of making no repairs at all if it is less costly to pay the fee.

We use data from the California I/M Review Committee and from the Sun Oil Co. study to estimate repair effectiveness. Emissions after repair are a function of pre-repair emissions, age of the vehicle and the cost of repair.

$$E_j^1 = \beta_0 + \sum \beta_{ij} E_i^0 + \gamma_1 Age + \gamma_2 Cost + \delta \tag{7.2}$$

where $i, j =$ HC, CO, NO_x are the pollutants of interest, E^0, E^1 refer to emissions before and after repair, respectively, *Age* is the vehicle age in years, *Cost* is the reported repair cost, and δ is the disturbance term, $\delta \in N(0, \sigma_r)$.

Estimation results are shown in the Appendix, Tables 7.A3 and 7.A4. Pre-repair emissions are the most significant predictor of after-repair emissions, and the cost of repair is not significant for most pollutants. The regression model is not a causal model of repair effectiveness; the results merely express associations between post-repair emissions and a few pieces of information known pre-repair. If anything, the results may underestimate the mechanic's ability to predict the emission reductions from repair. When looking at a specific vehicle, the mechanic will have more information and will presumably be able to predict repair effectiveness at least as well as the model predicts.

In the simulations described below, if cars are to be mended, we assume up to two stages of repair. The first will resemble the less costly and less effective California repairs, while the second will resemble the Sun Oil Company repairs. In both cases, some repairs involved only minor engine adjustments and were made at zero cost. Accordingly, we assume the cost of repair follows a compound distribution, in which repair costs are non-zero with probability p and non-zero repair costs observe a log normal distribution.

7.3.2. *Simulations*

Simulation of the command-and-control I/M programme

The I/M programme is characterized by a set of cutpoints, one for each of the three pollutants, HC, CO and NO_x, the function of which is to determine whether the vehicle passes the test. The simulation proceeds as follows. (1) Initial vehicle emissions measurement. For each vehicle, an emissions measurement is determined for each vehicle based on the error distribution from estimation of equation (7.1), shown in Table 7.A2. The measured emissions are the differences between the vehicle's true emissions and the measurement error. (2) Repair. Repair any vehicle that has measured emissions exceeding a cutpoint for any of the three pollutants. After-repair emissions are determined by applying the coefficients from estimation of repair effectiveness based on the California I/M study (Table 7.A3) plus a random error distributed as the errors in that table. The cost of the repair is a random draw from the empirically determined repair costs from the California I/M Review Committee study. (3) Retest. A second draw is made from the emission test distributions shown in Table 7.A2 and subtracted from the true post-repair emissions as determined in the preceding step to get the measured emissions. (4) Second repair. If any cutpoint is still exceeded, proceed to a second repair using coefficients, costs and errors from the Sun Oil Company study (Cebula, 1994) shown in Table 7.A4.

Simulation of the economic incentive programme

The emissions fees are charged against the emissions rate of the vehicle for each pollutant. Motorists can decide whether to pay the fees without repair or repair the vehicle and pay the resulting new fees. The simulation works as follows. (1) Initial vehicle emission measurement and calculation of initial fee.

$$Fee_0 = \max\left[0, \sum t_i(\tilde{E}_i^0 - Baseline_i)\right] \tag{7.3}$$

where t_i is the fee rate for each pollutant, $\tilde{E}_i^0$ the measured emission rate of pollutant i (as in the Command and Control policy above), and *Baseline* the level of free emissions granted each vehicle, if there is one. (2) Prediction of emissions reductions and estimation of post-repair fee. After repair emissions, $\tilde{E}_i^1$ are predicted using the California I/M study from equation (7.2) (coefficients shown in Table 7.A3). Estimated after-repair fees are determined as:

$$EstFee_1 = \max\left[0, \sum t_i(\tilde{E}_i^1 - Baseline_i)\right] \tag{7.4}$$

(3) Motorists decision rule is the following:

Repair if $Fee_0 > RepairCost + EstFee_1$.

(4) Repeat (2) and (3), using the Sun Oil Co. repair cost and effectiveness data (Cebula, 1994).

In comparing C&C policies and the emission fees described here, there may be a temptation to treat the fee policy baseline as an analogue of the cutpoints in the regulatory programme. In one sense this is reasonable and useful, since both trigger some kind of action by the authorities, either a requirement to repair the vehicle or a fee levy, some or all of which can be avoided by the repair of the vehicle. However, the correspondence can be pushed too far, for in the textbook treatment of these two policies the correspondence does not exist. In the textbook, in fact, fee baselines do not exist, and the appropriate correspondence is between emission quantity limits in the C&C programme and emission fee rates.[16] Real policies are more complicated, and the cutpoint or baseline is only one of several parameters affecting the stringency of the policy.

Assumptions behind the simulations

The simulation model must make a number of simplifying assumptions which deserve some explanation. Among the technical assumptions, probably the most important concern the structure of the repair submodel. As noted above, repair is allowed in two stages, but at each stage only one repair option is considered, whereas mechanics will often have numerous options, each with its own emission reduction implications. Unfortunately, the repair data that would allow the estimation of such a structural model are unavailable. As far as we are aware, all repair data sets provide emission data only after all repairs have been made. A structural model would require estimates of the emission effects of several alternative repairs or repair combinations. With this limitation, we believe that our two-step model approximates what mechanics actually do when faced with uncertainty about repair effects and with motorists interested primarily in minimizing repair costs rather than in reducing emissions. They will do an inexpensive repair first, and see what the effects are before moving on to more costly repairs.

In simulating the operation of emission fee programmes we make a number of additional assumptions that warrant attention. First, we assume there is no principal-agent problem between mechanics and motorists. Mechanics generally have better information about vehicle performance and the cost of repair than do consumers. Indeed, there is some reason to believe that the principal-agent problem is more difficult for emission repair than for ordinary vehicle repairs, because motorists have more difficulty in observing whether emission repairs have been undertaken correctly. However, this problem is not unique to emission fees; it also affects C&C regulatory programmes. The limited evidence on repair in I/M programmes, however, does not indicate a cause for concern on this score. Indeed, one might say that the problem with those programmes is the reverse, that the mechanic may have identified too strongly with the consumer – and both are allied against the public good (see Hubbard, 1995).

The second assumption is of particular importance to the comparison of emissions fee policies and C&C policies: it is that motorists are risk-neutral. The evidence suggests that the benefits of emission repair are highly uncertain *ex ante*. In a regulatory programme where motorists with sufficiently high emissions are required to repair their vehicles, their risk aversion does not matter. If, on the other hand, motorists have a choice between paying known high fees versus paying repair bills that may reduce the fee but by a highly uncertain amount, they may be more likely to choose the less risky alternative even though in expectation they would be better off repairing their vehicles. However, there are several reasons why risk aversion may not be an important problem in this case. First, even very high estimated repair costs are small relative to the income of most individuals, so it is unlikely that risk aversion, at least as it is commonly represented in economic models, will make much difference in motorists decisions. Second, even if it became evident that risk aversion does affect repair decisions, markets are likely to take care of the problem. For example, it would be in the interest of mechanics to offer guarantees on the maximum fee liabilities for motorists.

The final and perhaps most important assumption of the model is that there is no fraud or evasion by mechanics or motorists. As discussed above, fraud has been a continual problem for I/M programmes and would be likely to be a problem for fees as well. Indeed, it is hard to think of any monitoring method that is not vulnerable in some way to evasion, whether by tampering or other means. However, just because a monitoring approach can be evaded does not mean it should be ruled out. The more relevant questions are, how common is it likely to be and how difficult or expensive would it be to prevent?

A more extensive discussion of fraud is beyond the scope of this paper. What is relevant here is whether considerations of fraud affect the choice of economic incentive approaches versus command and control. Under both fees and C&C policies, substantial penalties for gross emissions are possible, creating a strong temptation to engage in evasive behaviour for some motorists.[17] As we will see below, even though overall social costs of fees are generally lower, they often inflict higher costs on individual motorists, providing generally higher incentives to cheat. On the other hand, fees can more easily be based on the average of a number of test results, whether several lane tests or a number of remote sensing readings. To show compliance with a C&C programme generally requires only one passing test, no matter how many previous tests have failed. Tampering will be easier to accomplish if only one success is required. These are important issues for reducing vehicle emissions, but we do not attempt to model them in this version of the simulation, and focus instead on the problems of uncertainty and distribution of costs.

7.3.3. *Variation in technical parameters*

As observed in the preceding section, neither the accuracy of the emission measurements nor the repair-effectiveness predictions is very high. We simulate the improvement in such measurements by allowing a fraction of the unexplained variance in equations (7.1) and (7.2) to be 'explained', thereby reducing measurement error. Thus the emission measurement equation becomes:

$$\text{actual emissions} = \text{measured emissions} + (1 - \kappa)\varepsilon^* \qquad (7.5)$$

where κ, $0 \leq \kappa \leq 1$, is the fraction of the BAR-90 errors that we now assume can be explained, and $Var(\varepsilon^*) = Var(\varepsilon)$. As κ increases from 0 to 1, emission measurement error, $(1 - \kappa)\varepsilon^*$, decreases from that of the two-speed idle test to zero error.

Similarly, to model the improvement in repair prediction, equation (7.2) becomes

$$E_j^1 = \beta_0 + \sum \beta_i E_i^0 + \gamma_2 Cost + \lambda d + (1 - \lambda)\delta \qquad (7.6)$$

where $0 \leq \lambda \leq 1$ and $Var(\delta^*) = Var(\sigma)$. As λ increases from 0 to 1, repair predictions increase in precision from the nominal levels determined from the California and Sun Oil Company data to perfect predictions.

7.3.4. *Variation in policy parameters*

Command and control programme

We model two variables affecting the stringency and effectiveness of the C&C approach to I/M. The cutpoints are the levels of measured emissions that trigger a decision to require the vehicle to be repaired. The lower the cutpoints, the more vehicles will be repaired, and the greater will be the emissions reductions. Lower cutpoints also increase the probability of repairing the 'wrong' vehicles, that is, those with emissions in compliance or with marginally high emissions that are difficult to reduce. In most jurisdictions, including California, cutpoints are vintage-specific. The initial cutpoints are shown in Table 7.A5.[18]

The other policy variable that affects the performance of the regulatory programme is the maximum repair cost, or the maximum motorists are required to pay to repair their vehicles. If this limit is reached, then the motorist receives a waiver that allows the vehicle to be operated even though it may not meet emission criteria. As we suggested above, however, the Clean Air Act Amendments (CAA) of 1990 required that states set repair waiver limits no lower than $450 per car for the enhanced I/M policy.

Emission fee programme

The parameters specific to the emission fee programme are the per-unit fees (t_i in equations (7.3) and (7.4)), the zero-fee baselines, if any, the motorist's maximum obligation, and possible repair subsidies.

Per-unit fees. These are the rates in g/mile that motorists are charged for emissions. Clearly, the greater the fee rates per unit of emissions, the greater the emission reductions and the greater the motorists' repair costs and fee payments.

Zero-fee baselines. When baselines are used, motorists only pay fees on emissions above some threshold. As in the regulatory policy, the baselines may be vintage-dependent, which may be more equitable but is also likely to be less efficient than a constant baseline for all vehicles. The purpose of these baselines is to reduce the amount of fees collected, and hence minimize the redistributional impact of the fees. Higher baselines mean, in addition to reduced fee collections, fewer vehicles repaired and smaller emission reductions. To increase emissions reductions, then, one can choose either to reduce the baseline or to increase the unit fees.

Maximum obligations. In C&C I/M programmes waiver limits are imposed to limit the maximum that any motorist can spend. For emission fees the analogue is a limit on the total fee that motorists are required to pay. Thus, the motorist chooses whether to spend the maximum on fees or on a combination of test and repairs, depending on which is least expensive. If both fees and repair costs plus after-repair fees are greater than the maximum expenditure limit, it is not clear which option the motorist will choose. On the one hand, there may be some non-pecuniary costs of repair that would incline the motorist to pay the fee. On the other, the motorist may choose repair because the cap distorts the motorist's valuation of the outcomes. Suppose, for example, the fee on the initial test is $500 and the fee plus repair cost limit is $450. If the repair cost is $250 and the expected post-repair fee is $300, then the repair is not cost-effective. Nonetheless, the motorist may undertake this repair because, in both cases the expected obligation exceeds the maximum of $450. By repairing the vehicle the motorist cannot do worse than paying the $450, but if the repair is more successful than anticipated, i.e. it results in post-repair fees of $200 or less, he can possibly do better. The motorist may make this repair, up to the spending maximum, if the potential gain exceeds the inconvenience of undergoing repair. We assume in the simulations that follow that the motorist will make the repair when both the fee before repair and the repair cost plus after repair fee exceed the maximum obligation.

Repair subsidies. In principle, repair subsidies are one way of assuring that desirable repairs are made. The idea is to use the collected emission fee revenues to make cost-effective repairs on vehicles for which the owner has paid up to some limit. By making the state responsible for repair costs beyond the limit that can be imposed on individual owners, the emissions reductions achieved are the same as if there had been no limit on individual expenditure.

Several practical considerations should be mentioned. First, there is no guarantee that the sum of emission fees collected will exceed the sum of subsidies needed. Fee collections are especially influenced by the size of the zero-fee baselines, and so this parameter can affect whether fee revenues will be sufficient. Second, since the state makes the decision on whether certain repairs are made some, possibly cumbersome, administrative machinery may be necessary. Finally, any subsidy policy is always subject to incentive problems. Motorists or mechanics may try to ask for more repairs than are necessary to obtain the subsidy.

7.4. Results

The results of the simulation are presented for a 1000-car fleet with emissions characteristics of the California fleet (see Table 7.A1). We first compare the costs and benefits of emissions fee policies with the C&C policies under different assumptions for the technical or policy parameters. To compute benefits the true emission reductions are valued at \$10 000/ton for NO_x, and \$3000/ton for HC,[19] and in the figures discussed below we express emission reductions in terms of 'weighted tons', with weights of 0.3 for HC and 1.0 for NO_x. Thus, the monetary value of a weighted ton for the purpose of calculating benefits is \$10 000. Before-repair emissions in the simulated fleet total 30.6 weighted tons per year. The policies we examine reduce weighted emissions by about 5 to 11 tons in this fleet (15–35%). In the policy comparisons we keep track of both the social costs, which include the cost of inspection and repair (and not fees, which are transfers), and the motorists costs, which include inspection, repair and any fees the motorist must pay.[20] Net benefits of either policy are difference between total benefits and total costs at any level of emissions reductions.

To put the C&C results on an equal footing with the emissions fee, we have assumed that mechanics will not make repairs under C&C that would result in an increase in the expected weighted sum of the emissions. This allows us to make an optimistic but reasonable assumption about mechanic behaviour with respect to the I/M regulatory programme.

7.4.1. *Command-and-control policy vs emissions fees under uncertainty*

Figure 7.1a and b show the total costs and benefits of the C&C (I/M) policy and the emissions fee policy when both policies have a cap on the total amount any motorist must pay of \$450. In the case of the C&C policy, the \$450 is a limit on the total repair costs, and is mandated by the Clean Air Act. We include the \$450 cap for the fee as well. Under the fee cap, motorists are not required to pay more than \$450 in fees. Costs are shown for different levels of the emissions fee when λ (repair forecast accuracy) and κ (emission measurement accuracy) are either 0 to represent current uncertainty or 1 to represent perfect certainty.

Figure 7.1a shows the case of perfect certainty about emissions measurement

(a)

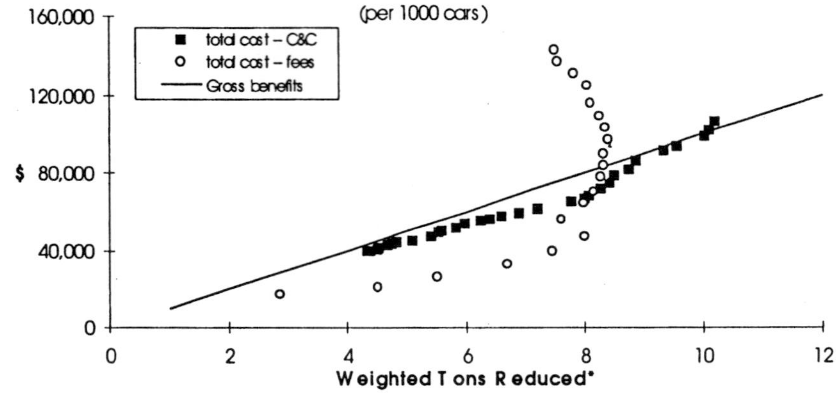

(b)

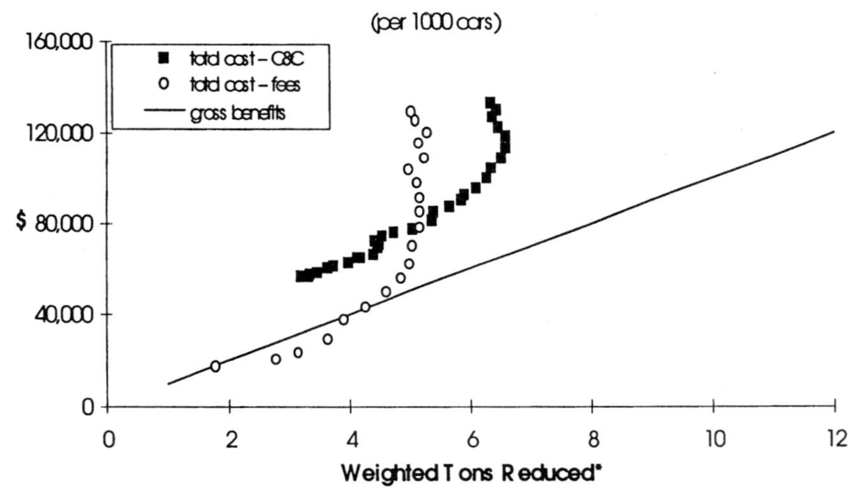

Figure 7.1. Comparison of C&C and fee programmes with caps on maximum payments by motorists. (a) Certainty case; (b) uncertainty case.
*Weights are 0.3 for HC and 1 for NO_x.

and repair predictability. Net benefits with the fee are quite a bit higher than the net benefits of the C&C policy up to about 8 tons reduced. After 8 tons, no further cost-effective repairs can be done without bumping into the cap, and it is impossible to reduce emissions further. Because fees must be paid in addition to repair costs, the emission reductions reach a maximum under a fee with a cap at lower emissions reductions than under C&C with the same cap.

Under the fee, net benefits (the vertical difference between the gross benefits line and costs) reach a maximum at about 7 tons reduced. The C&C policy in Figure 1a shows total costs to be close to gross benefits throughout the range of emissions reductions shown. Net benefits increase slightly and then decrease as emissions reductions increase (as the cutpoints are made more strict).

Under uncertainty (Figure 7.1b) both policies do much worse. The costs are much higher for both, with increasing marginal costs which eventually become very inelastic. The fee policy reaches the limit of potential emissions reductions at about 5 tons, the C&C policy at about 7 tons. Under uncertainty for both policies, attempts to reduce emissions beyond these levels would simply drive up costs. At low emissions level the emissions fee still has much lower costs than the C&C programme, and still shows slightly positive net benefits. Net benefits are negative at all levels of emissions reductions under C&C.

Figure 7.2a and b compares the C&C programmes and emissions fees policies when neither includes the $450 motorist expenditure cap. In this case it is even more clear that the net benefits of fees are greater than for the C&C policy, at any level of targeted emissions reductions. Uncertainty again affects both policies, but seems to have a relatively greater impact on the C&C policy. There are increasing marginal costs for both policies, especially under uncertainty, but again, as shown in Figure 7.1, only the fee policy shows positive net benefits under the uncertainty case. Under uncertainty, the C&C policy still shows steeply increasing costs at about 7 tons, even without a cap. Cutpoints set too tightly can result in fewer emission reductions and high costs, because they require the repair of vehicles whose emissions, under uncertainty, cannot be easily diagnosed and repaired. This outcome reflects the actual experience in California, where many of the marginally polluting vehicles reported in the California I/M Review Committee study (1993) had higher emissions after repair than they had before. Under current levels of uncertainty, whether using a cap on motorists expenditures or not, trying to set cutpoints in an I/M programme too strictly may only result in higher costs.

Comparison of Figures 7.1 and 7.2 suggests that there is a substantial loss of economic efficiency when limits are put on motorist expenditure, whether in a fee or a C&C programme. Another policy to reduce motorist obligation is the use of baselines in a fee policy, in which motorists do not have to pay unless their emissions exceed some level, in this case the baseline is set at the average emissions for all vehicles for each pollutant. Figure 7.3 shows the driver costs with and without a baseline fee, both with and without caps on the total motorist obligation. The driver costs include the social costs depicted above and the costs of the fees, which are a real cost to individual motorists.

With no cap, motorists costs rise steeply as the fee is raised, with over 85% of motorist costs in fee payments. With the baseline, costs to the motorist are typically reduced by 60–70%. Of course, some motorists with cars which are expensive to mend would still pay large fees under these baselines, but on average, motorist costs with the baseline are much lower than under the

(a)

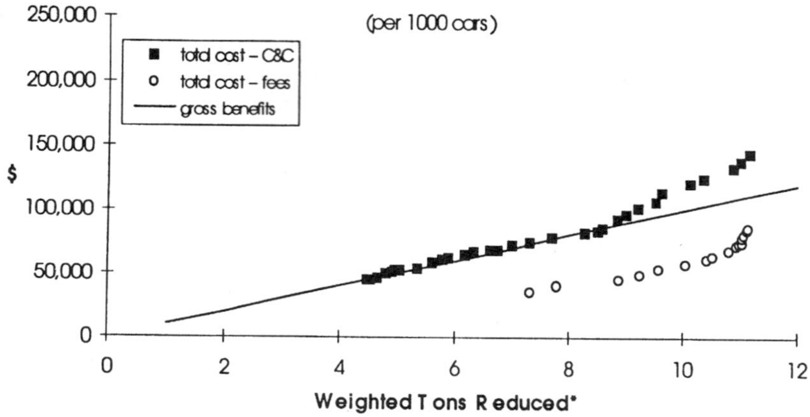

(b)

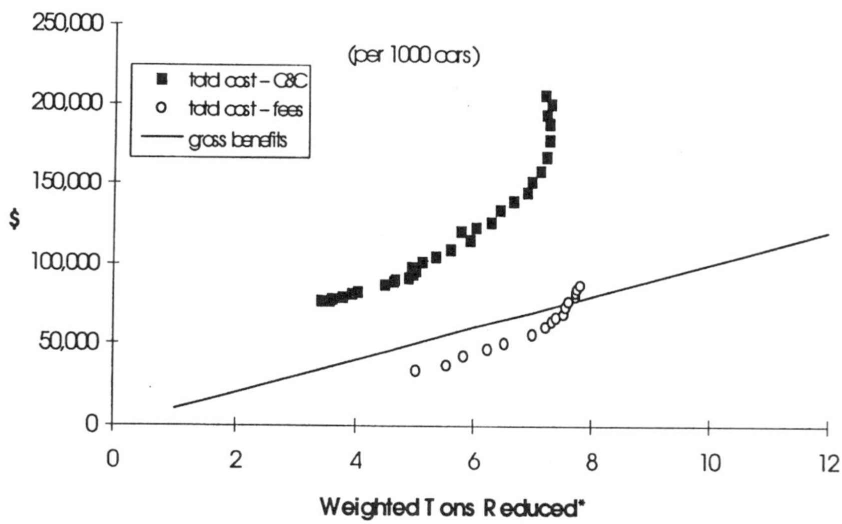

Figure 7.2. Comparison of C&C and fee programmes without caps on maximum payments by motorists. (a) Certainty case; (b) uncertainty case.
*Weights are 0.3 for HC and 1 for NO_x.

no-baseline case (Figure 7.3). The inclusion of a baseline reduces total driver costs dramatically, both when an emissions cap is present and when it is not. Although both caps and baselines lower motorists obligations, they do so in different ways. Caps primarily limit the motorist's maximum obligation, while baselines tend to reduce the average obligation.

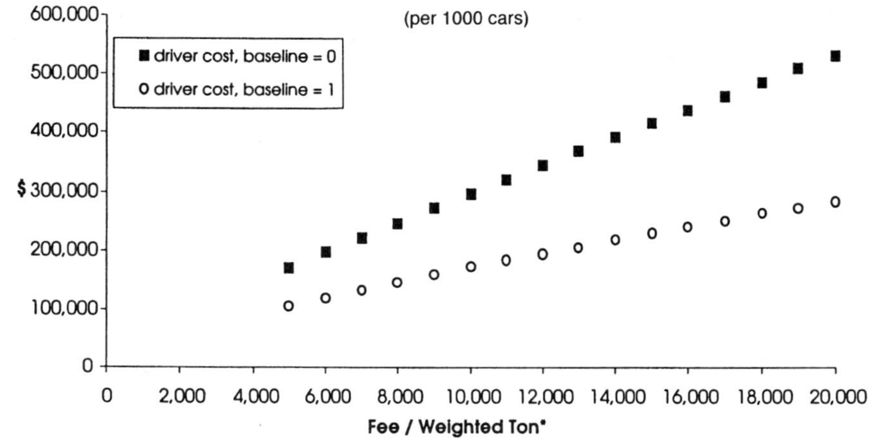

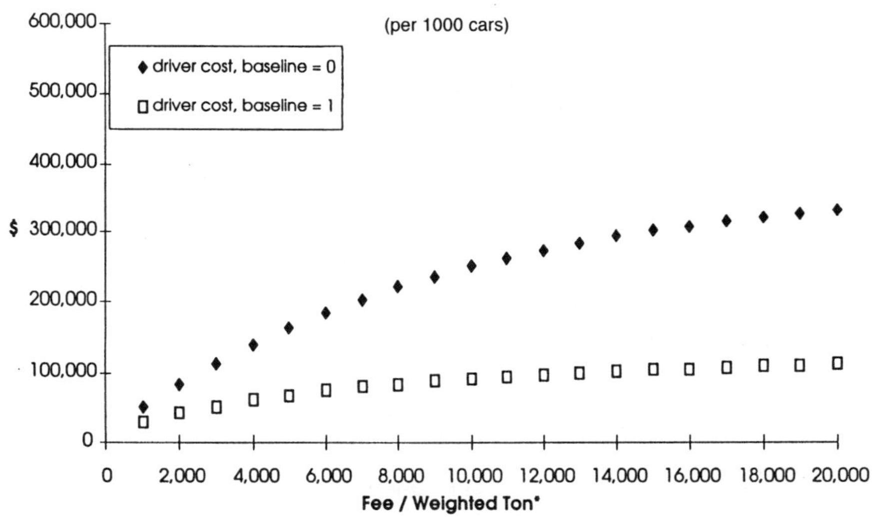

Figure 7.3. Effects of baselines on motorists' costs (driver costs include fees paid in addition to repair and inspection costs) with and without emission caps. (a) Without cap; (b) $450 cap.
*Weights are 0.3 for HC and 1 for NO_x.

These two policies are also likely to have very different efficiency results. Figure 7.4 compares the costs and emission reductions of fees with caps only, fees with baselines only, and fees with neither. Under certainty, the three policies all have similar costs at low levels of emissions reductions, but neither the fees with caps nor fees with baseline can achieve the higher emissions reductions obtained by a pure fee. The outcome is very different under uncertainty. Unlike the other two policies examined, the emission reductions produced by the fee

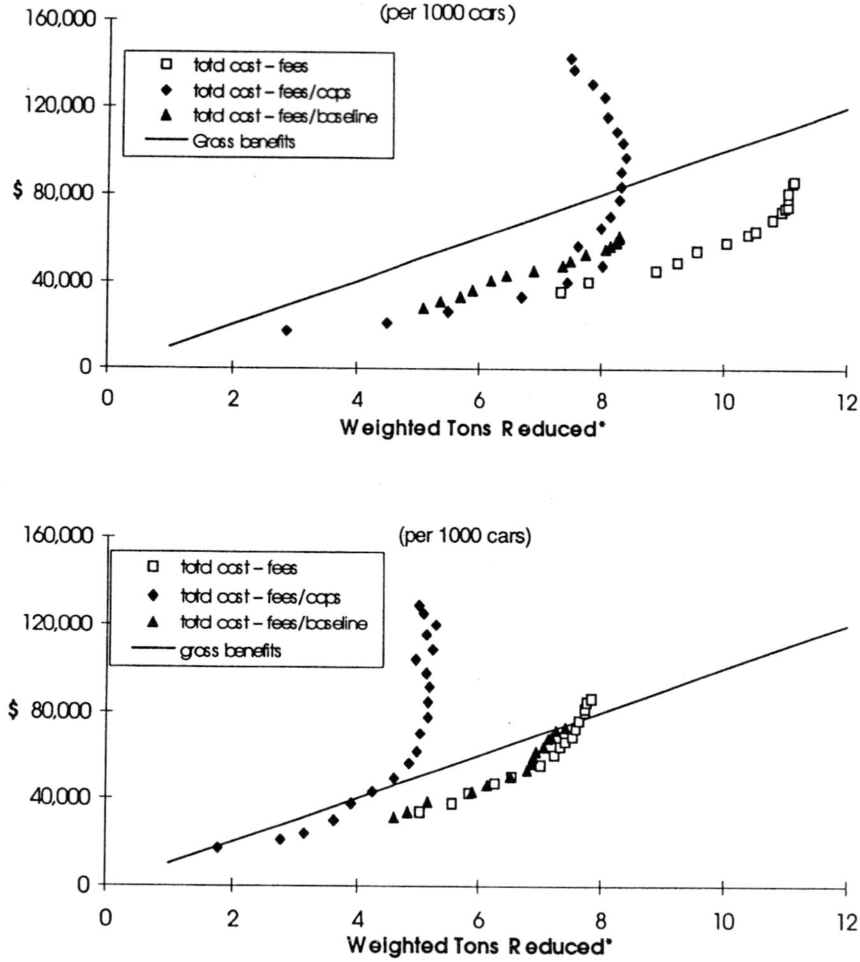

Figure 7.4. Comparison of pure fees, fees with caps, and fees with baselines. (a) Certainty case; (b) uncertainty case.
*Weights are 0.3 for HC and 1 for NO_x.

with baseline are but little affected by uncertainty. Its costs are very close to those of the pure fee for the same emission reductions. Both are much more efficient and can achieve higher emission reductions than the fee with caps. With current uncertainty, allowing a baseline with an emissions fee appears to be a good policy because it offers some distributional advantages and still attains about the same efficiency level as a pure fee.

Figure 7.5 compares the net benefits of the pure fee and the fee with the baseline as a function of the fee rate. We would expect from welfare theory (e.g., Baumol and Oates, 1988) that the introduction of a baseline would reduce

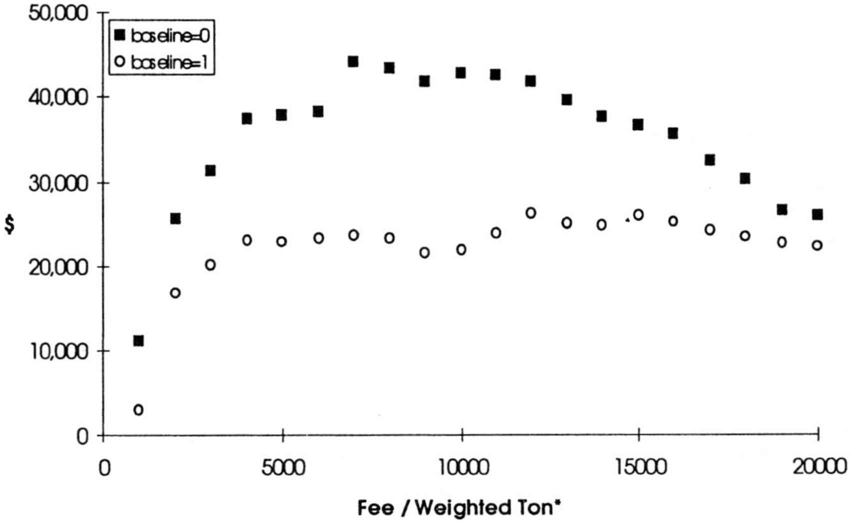

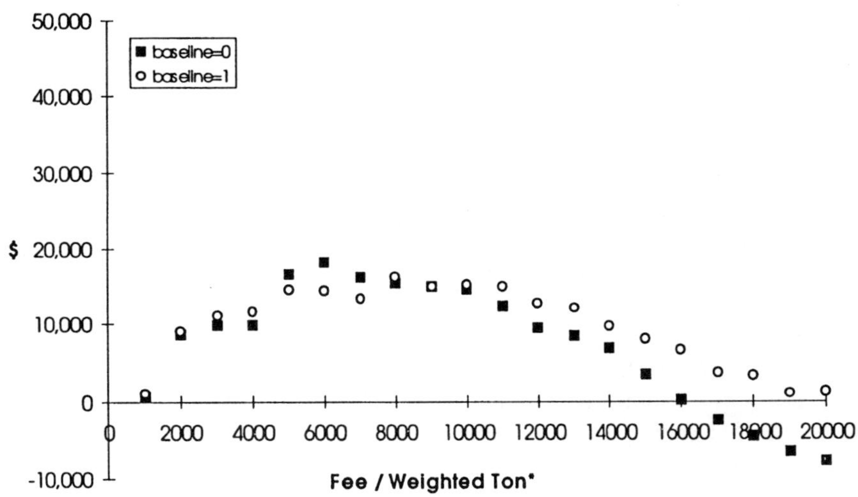

Figure 7.5. Net benefits with and without baseline. (a) Certainty case; (b) uncertainty case. *Weights are 0.3 for HC and 1 for NO_x.

the net benefits of an emissions fee policy, because some emissions are untaxed. This is the case under perfect certainty (Figure 7.5a) where the no-baseline fee has substantially higher net benefits at all fee levels. Under uncertainty, however, the results can be quite different (Figure 7.5b). Both policies have similar net benefits at all fee levels up to about $10 000/weighted ton, and at higher fees the fee with baseline has higher net benefits. Net benefits decline more rapidly under the no-baseline case as fee rates increase. Compared with a pure fee, it is at least possible that at some positive baseline fee, the net benefits of making fewer mistakes on the low emitting cars exceed the net benefits of missing some that should have been repaired. In this case, net benefits would be higher at a baseline above zero.[21]

7.4.2. *Combining fees with subsidies*

As noted above, caps or limits on total motorist costs, whether in an economic-incentive programme or in a regulatory programme, severely reduce the programme's effectiveness and efficiency. Such limits, unfortunately, prevent some cost-effective repairs from being made, and in the emission fee programme they may also encourage some repairs that are not cost-effective. One possible solution to this problem is to combine a cap on total motorist costs with a repair subsidy financed by fees collected on other vehicles. Such a cap would work like this: Suppose the fee before and estimated fee after repair are tE^0 and $t\tilde{E}^1$, respectively, the repair cost is R, and the cost cap is L. The motorist's payment is $M = \min(tE^0, t\tilde{E}^1 + R, L)$. For cars for which $M = L$ the car should be repaired if $tE^0 > tE^1 + R > L$. In that case, if $R < L$, the car is repaired and the motorist pays a fee of $L - R$. If $R > L$, the motorist pays L toward the repair and the remainder $R - L$ is subsidised out of fees collected on other cars.

The limit L should not be so high that it does not really provide much help to motorists, nor should it be set so low that the collected fees will not be adequate to permit all vehicles with excess repair costs to be repaired. It turns out that the repair ceiling that equates the fees taken in to the subsidies paid out is surprisingly low. Figure 7.6a and b shows fees collected and subsidies paid out when the base fee rate is \$10 000/weighted ton for baselines of 0 and 1. When the baseline is zero, the break-even ceiling is less than \$40. Raising the fee (not shown) raises the break-even ceiling (to about \$70 when the fee is \$20 000/weighted ton), and raising the baseline raises it even more. When the baseline is equal the mean emission rate and the fee is \$10 000/weighted ton, the break-even ceiling is about \$100.

If a way can be found to provide the correct incentives for vehicle repair, then such a policy should achieve the same emission reductions and costs as a fee without subsidies. The difficult question is, as with any subsidy policy, whether a procedure can be devised that would work in the real world, that is, would induce the right repairs to be made without either cumbersome bureaucracy or excessive fraud. The most obvious problem is that when the constraint on motorist sacrifice is active, the motorist is indifferent between repair and no repair, for either way his payment is L. This means the repair decision must be made by someone else. The mechanic is not a promising choice, which suggests that repair decisions must be passed on by some authority who can be counted on to make the cost-effective choice. That this is unlikely to be a trivial concern can be seen in Figure 7.7, which shows the

(a)

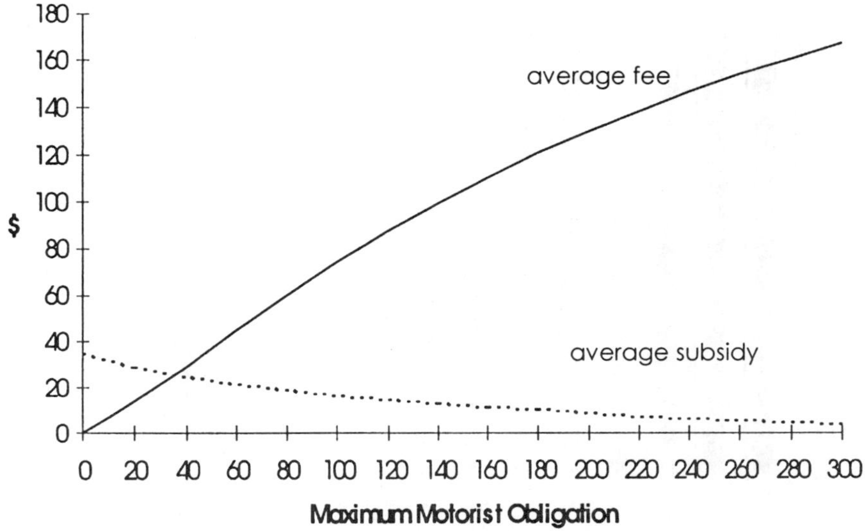

(b)

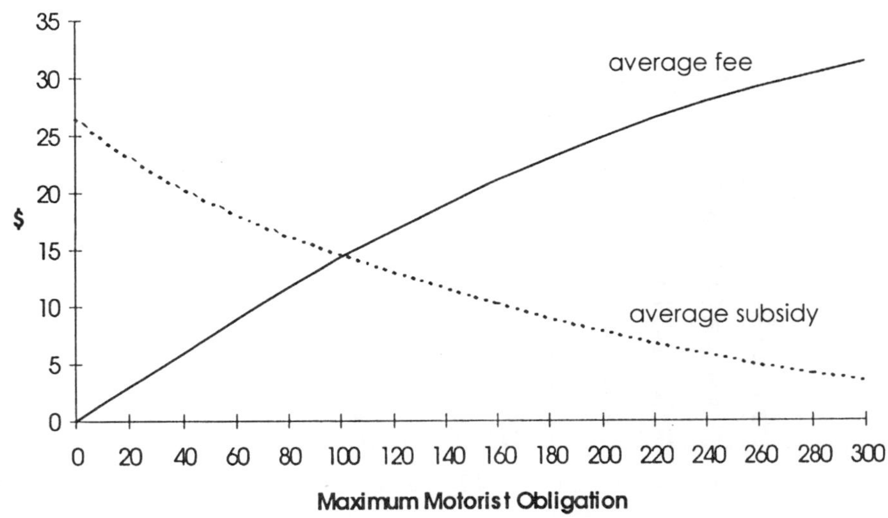

Figure 7.6. Average fees and subsidies with no baseline (a) and with baseline (b).

(a)

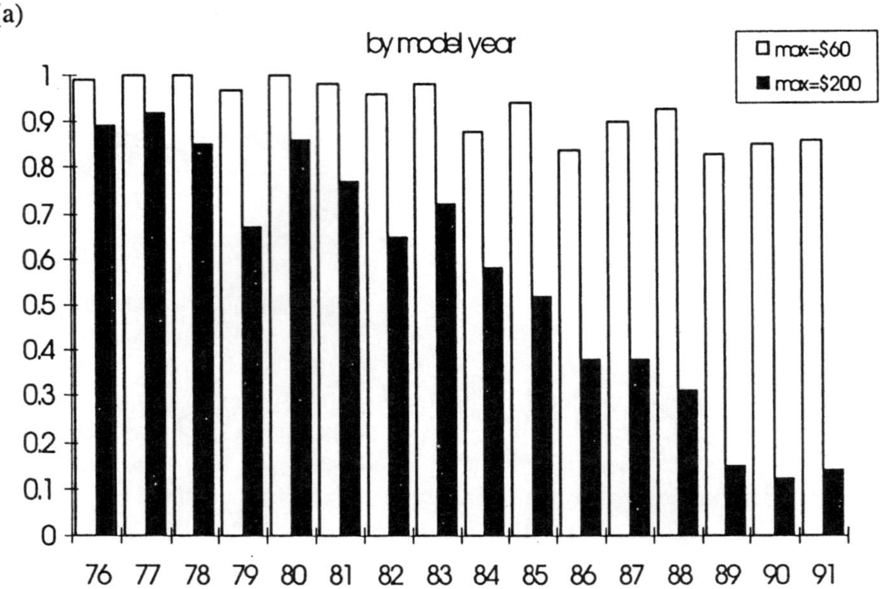

(b)

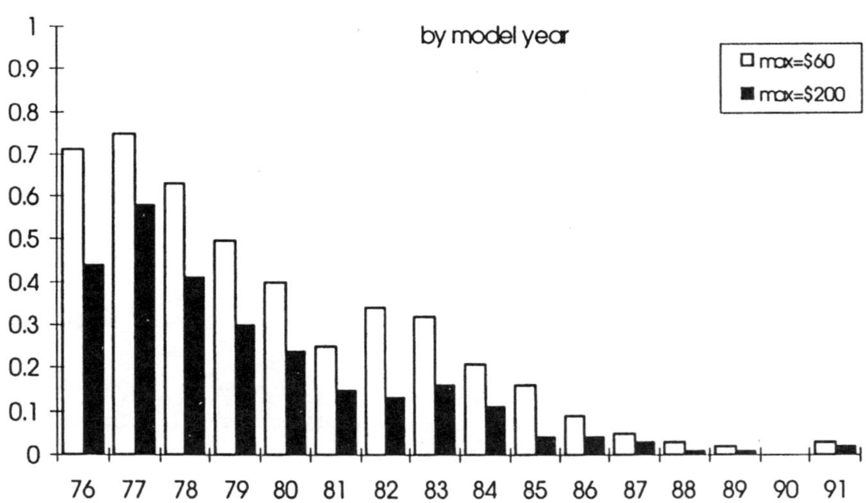

Figure 7.7. Proportion of vehicles paying the maximum with no baseline (a) and with baseline (b).

fraction of vehicles (by model year) paying the maximum in fees or repairs for ceilings of $60 and $200. At the $60 ceiling nearly all motorists pay the maximum, and at $200 nearly all the owners of older cars do. Fraud is also a potential problem, for there is both the temptation and opportunity for collusion between motorists and mechanics to find ways to treat ordinary maintenance as an environmental repair and hence eligible for subsidy.

7.5. Conclusions

We have compared a traditional regulatory programme for reducing vehicle emissions, so-called I/M programmes, with various emissions fee policies. We find that fees have higher net benefits than the C&C policy in all cases. However, we find that the extent of uncertainty in measuring and repairing vehicle emissions can have important impacts on the evaluation of policy alternatives. With uncertainty, both C&C and emissions fee policies have much lower potential to reduce emissions and have lower net benefits. The C&C policy is particularly affected by uncertainty, and there is a limit to the amount of possible emissions reductions. Setting cutpoints in the regulatory programme at too strict a level will result in high costs and little additional emissions reduction.

There are various policies that attempt to reduce costs of pollution control to motorists. Expenditure caps are used as part of the current I/M programme and can also be used with fees. The simulations found that caps reduce the efficiency of emissions fees substantially, especially when there is uncertainty. Caps also tend to set a limit on the total amount of emissions reductions that can be achieved under any policy.

An alternative to caps on expenditures which is explored in the paper is baseline fees. If motorists do not have to pay a fee until emissions exceed some baseline level, under certainty, net benefits are lower compared to a pure fee where motorists have to pay on all emissions. However, under uncertainty, the baseline fee is virtually no worse than the pure fee. The baseline introduces an error: some vehicles that should be repaired are missed, but it reduces another kind of error that occurs with the no-baseline fee when there is uncertainty since not as many low emitting vehicles will be repaired unnecessarily. Under current levels of uncertainty, a baseline fee set at average fleet emissions both reduces motorists costs, and preserves the efficiency of pure fees.

Even under a baseline fee, some motorists will have to pay large amounts to have their vehicles repaired. Another policy to reduce this burden is to allow caps or limits on total expenditures by motorists, but to offer repair subsidies financed by fees collected on other vehicles. The repair ceiling that equates the fees taken in to the subsidies paid out is surprisingly low, less than $100 even under uncertainty. This has promise as a way of distributing the costs of vehicle emissions repairs more equitably, but may face difficult incentive and implementation problems.

Table 7.A1. Average and total emissions of simulation fleet (average mileage 12 190 miles/year).

	Average emissions (g/mile)	Total emissions (tons/year/1000 vehicles)
HC	2.08	25.3
CO	30.7	374
NO_x	1.88	23.0

Appendix

7.A1. *Fleet emissions*

Table 7.A1 shows emissions characteristics of the 1000-car fleet used in the simulations, and created by making a series of random draws. First, vehicle vintages were drawn from a distribution reflecting the actual pattern of vintages in the California fleet, as determined by the 1990 NPTS. For each vehicle, an annual mileage was determined by a random draw from a vintage-specific mileage distribution, and emissions were randomly drawn from vintage-specific pollutant distributions. As described in the text, HC and CO are assumed jointly distributed, based on the vintage specific emissions distributions of vehicles in the Los Angeles region as measured by remote sensing of over 90 000 vehicles in 1991. The distribution of NO_x emissions was taken from a sample of 7234 vehicles given IM240 tests at EPA's Hammond, Indiana test facility. NO_x emissions are assumed independent of HC and CO. The emissions of the three pollutants were assumed to be lognormally distributed. The annual mileage was assumed to be log-normally distributed and independent of emission rates.

7.A2. *Test accuracy*

Table 7.A2 gives the results of estimation of equation (7.1). The California Bar-90 test is a no-load two speed test, measuring HC and CO at both idle and high speed idle at 2500 rpm. The HC idle test results are in parts per million and CO idle test results are in percent CO. FTP emissions are measured in g/mile.

7.A3. *Repair model estimation*

The results for California shown in Table 7.A3 indicate that by far the most significant predictor of post-repair emissions of a pollutant are the pre-repair emissions of that same pollutant. The coefficients show the marginal effectiveness of repair at removing pollutants from vehicle emissions in the California programme. For example, the coefficient of 0.36 for HC means that repair

Table 7.A2. Ability of the idle and high speed tests to predict FTP emissions (g/mile).

	FTP HC	FTP CO	FTP NO_x
Constant	1.235	21.287	2.218
	(0.308)	(1.750)	(0.0708)
Idle HC	0.00629	0.00325	0.777e-4
	(0.00063)	(0.0036)	(0.00014)
High speed idle HC	0.00481	0.000764	0.00068
	(0.00073)	(0.00416)	(0.00017)
Idle CO	0.292	7.512	0.026
	(0.093)	(0.528)	(0.0213)
High speed idle CO	0.0554	7.694	−0.217
	(0.1107)	(0.6299)	(0.026)
R-square	0.39	0.48	0.12
Standard error	5.67	32.24	1.30
N	669	669	669

Source: California I/M Review Committee, 1993. Standard errors in parentheses.

Table 7.A3. Predicting post-repair emissions (g/mile): California I/M review.

	HC	CO	NO_x
Constant	0.041	5.30	0.14
	(0.25)	(1.40)	(0.048)
HC_0	0.36	0.53	−0.0025
	(0.027)	(0.15)	(0.0051)
CO_0	0.026	0.62	0.0027
	(0.0045)	(0.024)	(0.00084)
NO_{x0}	0.32	0.60	0.73
	(0.13)	(0.72)	(0.025)
Cost	−0.00084	−0.011	−0.00027
	(0.00064)	(0.0034)	(0.00012)
Std. error	2.87	15.54	0.54
R-square	0.34	0.59	0.57
n	669	669	669

would remove 64% of additional HC emissions. The effects of other pollutants are not consistently related to post-repair emissions. Cost is significant for CO and NO_x and has the correct sign for all three pollutants, but in all cases the numerical magnitudes are so small that it has no practical importance (for example, expenditure of \$100 reduces expected HC emissions by 0.1 g/mile). Age is significant in the CO and HC equations, and the sign suggests either that older vehicles are more difficult to repair or that what must be considered the emission level for a fully repaired vehicle increases slowly with vehicle age. On average, a year of age increases post-repair emissions by 0.1 g/mile for HC

Table 7.A4. Predicting post-repair emissions (g/mile): Sun Oil Company.

	HC	CO	NO_x
Constant	0.36	2.48	1.04
	(0.28)	(3.45)	(0.24)
HC_0	0.098	0.16	0.10
	(0.026)	(0.32)	(0.023)
CO_0	−0.0036	−0.00051	−0.0041
	(0.0015)	(0.018)	(0.0013)
NO_{x0}	0.0055	−0.26	0.011
	(0.020)	(0.24)	(0.017)
Cost	0.0013	0.023	0.00017
	(0.00075)	(0.009)	(0.00066)
R-square	0.11	0.05	0.12
Std. error	1.20	14.58	1.05
n	151	151	151

Table 7.A5. California emission cutpoints (g/mile).

Vehicle model year	HC	CO	NO_x
1972 and before	6.8	39.0	3.6
1973–74	6.4	39.0	2.0
1975–76	2.7	18.0	2.0
1977–79	1.23	18.0	1.5
1980	1.23	18.0	1.0
1981–88	0.6	10.5	1.0
1989+	0.39	7.0	0.4

Source: Klausmeier, R., 1995, 1995, Evaluation of the California Pilot Inspection/Maintenance (I/M) Program.

and 1 g/mile for CO. Again, these effects are relatively trivial compared to the variation in individual vehicles.

The Sun Co. data (Table 7.A4) show even higher pollutant removal efficiencies, owing most likely to the greater repair expenditure. Repair cost was insignificant and has been omitted from the model in the table.

The coefficients of the linear model may be biased because the dependent variable is truncated at zero, and a more searching analysis of the determinants of repair effectiveness would consider this. Such an analysis is beyond the scope of this paper, and in any event would be more usefully carried out with data on the specific repairs performed. In the simulation model described in the next section we use the linear model because its coefficients are easier to interpret and because we are more interested in prediction of the value of the dependent variable than in the coefficients of the independent variables.

Notes

1. An earlier draft of this paper was given at the Southern Economics Association Meetings, New Orleans, Louisiana, November 18–20, 1995. The authors would like to thank Trellis Green, Alan Krupnick, Robert Slott, Erik Verhoef and Margaret Walls for helpful comments on early drafts of this paper.
2. In a different application of an economic incentive programme for reducing vehicle emissions, Kling (1992) examines the California Low Emission Vehicle programme, which allows for permit trading for emissions control on new cars. She finds that the savings from the permit programme may be small relative to the regulatory programme because the regulatory programme was able to capture some of the cost savings of the incentive programme through requirements for smaller cars.
3. This system is similar to that examined by Roberts and Spence (1976), except that the baseline levels here are not tradable licenses.
4. Because there is no obvious source of revenue under a traditional I/M programme, as there is with an emission fee policy subsidies in a regulatory programme will require use of general revenues or new taxes.
5. EPA developed a technically sophisticated emission test protocol that included use of expensive automatic analysers and a dynamometer, called the IM240 test. This test is supposed to be more accurate and somewhat less susceptible to motorist evasion than the simpler idle tests, but it also more expensive.
6. The policy simulated is most like a centralized I/M system There are other ways to identify polluting vehicles, including remote sensing of tailpipe emissions. See Bishop and Stedman, 1994, Harrington and McConnell, 1993.
7. Vehicle scrappage in the context of this model is examined in Alberini et al., 1996.
8. A VMT-emission fee would in principle be more efficient than an emission rate fee. How much more efficient depends on the characteristics of the vehicle fleet and on the pollutants being controlled. For policies targeted at pollutants such as HC and CO, which vary a great deal from one vehicle to another, an emission rate fee achieves almost the same emission reductions as a VMT-emission fee. For policies targeted at NO_x, for which the variation across vehicles is less, the difference in results between the two policies is greater. In the current case, the net benefits of a VMT-emission fee exceed those of an emission rate fee by about 30%. The problem of fraud, which is a major problem for in-use emission problems generally, is much worse for VMT-emission fees because of the ease of manipulating odometers, particularly in older vehicles.
9. See McConnell and Harrington (1992), Aroesty et al. (1994) and Harrington and McConnell (1994) for more discussion of I/M issues.
10. The data and initial parameter values are discussed in more detail in the next section.
11. The EPA collected data from their Hammond, Indiana test lane over several years from 1991 to 1993. IM240 tests were performed on hundreds of vehicles for which there are HC, CO and NO_x readings. We used the variance–covariance matrix from this data to infer the relationship among pollutants for the remote sensing data.
12. The FTP test, or Federal Test Procedure, is an hour long test that includes the use of a dynamometer and the operation of a vehicle over many different driving modes, including acceleration, deceleration and stop and start. It is quite expensive to perform but is currently considered by regulators and the industry to be the best method for determining an estimate of actual vehicle emissions. Recently, however, the FTP cycle has been criticized for not accurately reflecting true driving behaviour and accompanying on-road emissions. It has been found to underrepresent high acceleration episodes (Ripberger et al., 1995). We use it as a measure of the true emissions because it is considered to be the best emissions measurement of the test data available. The IM240 test is simply a piece of the FTP cycle, and remote sensing data, which is likely to capture real world behaviour of drivers, has been criticized for its

accuracy (Smith, M.G., 1995; Glover and Clemmens, 1992) and because each reading capture only one phase of the driving cycle. We use the FTP to represent true emissions here, but we could use an alternative measure if one were available.

13. The Bar-90 test does not presume to measure NO_x emissions, but we nonetheless regressed the FTP NO_x measurements against the CO and HC results. It is no surprise the fit is so poor. Considering that any programme regulating NO_x emissions would have a way of measuring them, our results probably exaggerate the NO_x measurement errors.
14. The Bureau of Automotive Repair was originally established as a consumer protection agency, with the mission of fighting fraud in the auto repair industry.
15. Fully 40% of the vehicles were not brought into compliance even after expenditure of $450. It should also be pointed out that the vehicles were repaired by mechanics given special training.
16. Under highly restricted conditions (only one discharger or all dischargers with identical marginal cost functions) finding the optimum emission fees and optimum emission standards are dual problems.
17. Under VMT fees, where miles travelled matter in setting fees, the incentives to tamper would be even stronger.
18. The ratios of the h_i to one another reflect the relative stringency of the standards. For example, to remove regulation of NO_x, set h_3 to be a very large multiple of h_1. The cutpoints for California which we assume are shown in Table 7.A1.
19. See Small and Kazimi (1994) and Krupnick and Portney (1991). Since we are focusing only on the ambient ozone problem here and, in addition, the benefits of reducing CO are small relative to HC and NO_x reduction benefits, we have not included CO reduction benefits in this study.
20. The total cost of the inspection test itself (including motorist inconvenience) is between $10–20 per test per year (see McConnell and Harrington, 1992). In the simulations for this paper we use an average inspection cost of $15 a test.
21. This possibility can be demonstrated by a simple example. Suppose all cars in the fleet have emissions of 2 g/mile, 10 000 miles/year, or 0.02 tons/year. Each car can be repaired so as to eliminate its emissions for $250. Set the emission fee equal to the constant marginal damages of $10 000/ton. The benefits of repair are $200, so in a world of certainty no repairs should be made. Now suppose the measured emissions are 1 g/mile for half the cars and 3 g/mile for the other half. Although the emission measurement is unbiased in this world, half the cars would be repaired under a pure fee. For a baseline of anything greater than 1 g/mile, the fee with baseline would repair no cars, which is the welfare-maximizing outcome. Generally speaking, emission meaurement errors allow cars to be misclassified. Fees with baselines reduce the probability of false positives, while increasing the probability of false negatives. The optimum balancing of these two errors depends on the underlying true emission distribution and does not necessarily require the baseline to be zero. For the same reason, measurement uncertainty allows the optimum emission rate to be less than the marginal damages. Uncertainty about repair effectiveness can exacerbate, but can not by itself cause, this condition. Note the difference between this case and the case of abatement-cost uncertainty considered by Roberts and Spence (1976). They find that a mixed fee-license system (similar to our baseline) is optimal when damages are strongly nonlinear, but the pure fee is optimal when damages are linear. In this case it is possible for a fee with a non-zero baseline to outperform the pure fee even when damages are linear.

References

Alberini, A., W. Harrington and V. McConnell, 1996, Fleet turnover and old car scrap policies, paper presented at the American Economic Association Meetings, San Francisco, California, January.

Aroesty, J., L. Galway, L. Parker, M. Kamins, P. Wyn Wicinas, G. Farnsworth and D. Rubenson,

1994, Restructuring smog check: a policy synthesis, RAND Report No. DRU-885-CSTC, prepared for the California Senate Transportation Committee.

Baumol, W.J. and W. Oates, 1988, The Theory of Environmental Policy, second edition, Cambridge: Cambridge University Press.

Bishop, G.A. and D.H. Stedman, 1994, Remote sensing and vehicle variability, presented at the Fourth Annual CRC On-Road Vehicle Workshop, March.

California I/M Review Committee, 1993, Evaluation of the California smog check program and recommendations for program improvements, Fourth Annual Report to the Legislature, February 16.

Cebula, F.J., 1994, Report on the Sunoco Emissions Systems Repair Program, Philadelphia: Sun Oil Company.

Eskeland, G.S. and S. Devarajan, 1996, Taxing Bads by Taxing Goods: Pollution Control with Presumptive Charges, Washington, DC: The World Bank.

Fullerton, D., 1995, Why Have Separate Environmental Taxes? Working Paper, Austin, TX: University of Texas at Austin.

Glover, E.L. and W.B. Clemmens, 1992, Identifying excess emitters with a remote sensing device: a preliminary analysis, in I/M Costs, Benefits and Impacts Analysis, Ann Arbor, MI: USEPA, Office of Mobile Sources.

Harrington, W. and V.D. McConnell, 1993, Cost-effectiveness of remote sensing of vehicle emissions, in R.F. Kosobud, W.A. Testa and D.A. Hanson, eds., Cost-Effective Control of Urban Smog, Chicago, IL: Federal Reserve Bank of Chicago.

Harrington, W. and V.D. McConnell, 1994, Modeling in-use vehicle emissions and the effects of inspection and maintenance programs, Journal of Air and Waste Management, 44, 791–799.

Harrington, W., M. Walls and V.D. McConnell, 1995, Driving our way to cleaner air, Issues in Science and Technology, Winter 1994/1995.

Hubbard, T.N., 1995, Seeing Through the Smog: On Regulating In-Use Vehicle Emissions With Inspection and Maintenance Programs, UCLA Working Paper.

Kessler, J. and W. Schroeer, 1993, Meeting Mobility and Air Quality Goals: Strategies that Work, U.S. Environmental Protection Agency, Office of Policy Analysis, Final Draft, October.

Klausmeier, R., de la Torre Klausmeier Consulting, Inc., 1995, Evaluation of the California Pilot Inspection/Maintenance (I/M) Program, prepared for California Bureau of Automotive Repair, with Radian Corporation, Austin, Tx., March 31.

Kling, C., 1992, Emission Trading vs Rigid Regulations: The Robustness of Cost Savings Estimates for Vehicle Emissions, Working Paper, Davis, CA: University of California.

Krupnick, A.J. and P.R. Portney, 1991, Controlling urban air pollution: a benefit cost assessment, Science, 252, 522–528.

Lawson, D., 1993, "Passing the test" – human behavior and California's smog check program, Journal of Air and Waste Management, 43, 1567–1575.

Lodder, T. and K.B. Livo, 1994, Review and Analysis of the TOTAL Clean Cars Program, Denver, CO: Regional Air Quality Council and the Colorado Department of Public Health and Environment.

McConnell, V. and W. Harrington, 1992, Cost-Effectiveness of Enhanced Motor Vehicle Inspection and Maintenance Program. Quality of the Environment Discussion Paper QE92-18, Washington, DC: Resources for the Future.

National Research Council, 1991, Rethinking the Ozone Problem in Urban and Regional Air Pollution, Washington, DC: Committee on Tropospheric Ozone Formation and Measurement, National Academy Press.

Ripberger, C.T., M.O. Rodgers, P.O. Groblicki and J.P. Markey, 1995, Emissions from In-Use Vehicles During Closed-Loop but High-Speed Driving, presented at the Fifth CRC On-Road Vehicle Emissions Workshop, San Diego, CA, April 3-5.

Roberts, M.J. and M. Spence, 1976, Effluent charges and licenses under uncertainty, Journal of Public Economics, 5, 193–208.

Small, K.A. and C. Kazimi, 1995, On the costs of air pollution from motor vehicles, Journal of Transport Economics and Policy, 29, 7–32.

Smith, M.G., 1995, Key Issues Related to Regulatory Use of Remote Sensing, presented at the 5th CRC On-Road Vehicle Emissions Workshop, San Diego, CA, April 3–5.

Smith, S., 1995, 'Green' Taxes and Charges: Policy and Practice in Britain and Germany, London: Institute for Fiscal Studies.

US Environmental Protection Agency, 1981, Update on the Cost-Effectiveness of Inspection and Maintenance, Ann Arbor, MI, USEPA Office of Mobile Source Air Pollution Control.

Walls, M. and J. Hanson, 1995, Measuring the Incidence of an Environmental Tax Shift: The Case of Motor Vehicle Emissions Taxes, Draft Working Paper, Washington, DC: Resources for the Future.

White, L., 1982, US mobile source emissions regulation: the problems of implementation, Journal of Policy Studies, 11, 77–85.

CHAPTER 8

Forecasting the Environmental Effects of Road Pricing in London[1]

John Bates

8.1. Introduction

In 1991 the UK Department of Transport initiated a major research programme to assess the case for and against implementing 'congestion charging' on some basis within the Greater London area. Not only the costs and benefits were to be assessed but also the practical feasibility. The main results of the research programme are summarized in a three-volume Report (MVA, 1995). A vast amount of work was carried out, using a number of different models and testing and costing a large range of options. The assessment of the various alternatives involved many strands – economic, social, technological. This paper concentrates on a single aspect of the assessment, that related to environmental considerations. It is important, nevertheless, that these are seen in context.

Underlying the major aspects of assessment must be a system for forecasting how travellers will respond to a major change such as that brought about by road pricing. Within the overall research programme, the main forecasting model was known as APRIL (Bates et al., 1996). This was a so-called strategic model, representing the area of Greater London by means of 45 zones. It forecast the change in demand resulting from the road pricing charge, after taking account of supply-side effects, and was also able to calculate the resulting economic and environmental effects. The results in this paper are confined to those obtained from APRIL: however, it should be noted that other, more spatially detailed, models were also used in the overall assessment.

At the time of publishing the Report, the Secretary of State for Transport, Sir George Young, commented that the Government had "no current plans to introduce congestion charging, in London or anywhere else", and that the research had made clear "that congestion charging is not an option for the immediate future".

R. Roson and K.A. Small (eds.), Environment and Transport in Economic Modelling, 183–205.

Table 8.1. Trip purposes distinguished in APRIL.

Purpose	Home-based	Non-home-based
Work	HBW	–
Employers' business	HBEB	NHBEB
Education	HBEd	–
Other	HBO	NHBO

8.2. APRIL

APRIL is an incremental equilibrium model which predicts changes in travel resulting from a wide range of policies associated with congestion charging. It takes as given a base description of travel movements considered to be in equilibrium, and predicts changes relative to this base, given a change in generalized cost. The model can be broadly classified with a number of strategic models recently developed for urban transport studies (see Bates et al., 1991) in which much more emphasis is given to demand, and the spatial detail is reduced. Although the resulting model is clearly not able to give detailed network representation (and as noted above other models were used for this purpose), the challenge of such models is to include sufficient geographical information to allow the supply side relationships to be successfully approximated.

Demand is represented as a nested series of incremental logit models, dealing with various travel choices (for example: frequency, destination, mode, time of day), each with respect to a fixed base. The intention is to forecast medium-term responses to congestion charging, in which no consequential changes in car ownership levels or locations of residence and work occur.

Four home-based (HB) purposes, and two non-home-based (NHB) purposes are distinguished. These are shown in Table 8.1, together with their abbreviations, which are used in the text.

Within each home-based purpose, an average of seven person types is recognized, based on appropriate combinations of household income, household car ownership, and level of employer assisted motoring. The person types are, in general, specific to each HB purpose, and reflect the size of the market. Hence, for example, the level of employers' motoring subsidy only applies to HBW and HBEB. For NHB trips, no distinction between person types is recognized.

Four modes are recognized: car, bus, rail, and slow (walk and cycle). In addition, commercial vehicle (CV) movements are distinguished between light goods vehicles (LGVs) and all others (OGVs).

Seven time periods are used: pre-peak (06.00–07.00), morning (am) peak (07.00–10.00), morning (am) peak shoulder (10.00–11.00), mid-inter-peak (11.00–15.00), evening (pm) peak shoulder (15.00–16.00), evening (pm) peak

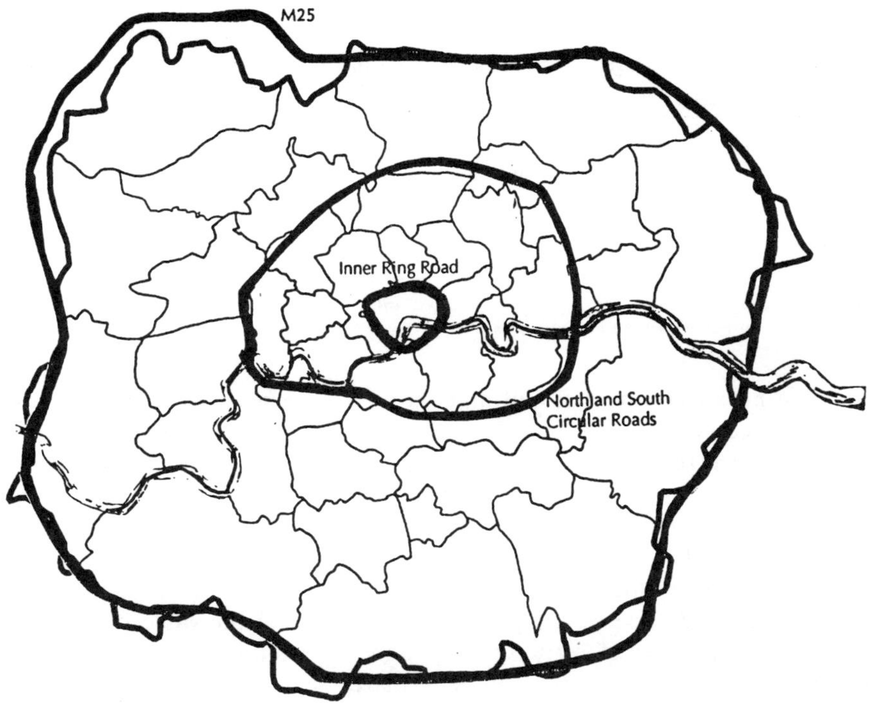

Figure 8.1. APRIL zoning system. ▬▬ Area boundary; —— zone boundary.

(16.00–19.00), evening (19.00–20.00). Provision is also made for journeys that return home after 20.00. Note that the number of weekday journeys commencing before 06.00 is insignificant. Only weekday traffic was represented, and no account has been taken of the possibility of travel shifting to the weekend.

APRIL has 45 internal zones representing broadly the area within the London orbital motorway (M25) (Figure 8.1), and a further seven external zones. The design of the system was based on an initial assessment of the most likely charging areas. In particular, it was conceived around the existing orbital routes in London – the inner ring road, which effectively defines the central area, the north and south circular roads, which define the boundary between Inner and Outer London, and the M25, which defines the boundary between outer London and the external area. If we treat these roads as approximate concentric circles, the radii are 3 km, 13.5 km and 28 km.

The total population living within the M25 is just over 7 million, of which two-thirds is in Outer London, and only 1% in Central London. On the employment side, there are nearly 4 million jobs, of which just over a quarter are in Central London, and 45% in Outer London.

The transport infrastructure is represented by a single integrated network. Different categories of roads or railways are represented by different link types,

Table 8.2. Allowable demand responses in APRIL.

Category of travel	Change route?	Change time?	Change mode?	Change destination?	Change frequency?
HBEd	yes	no	yes	no	no
HBEB	yes	no	yes	no	no
HBW	yes	yes	yes	no	no
HBO	yes	yes	yes	yes	yes
NHBEB	yes	no	no	no	no
NHBO	yes	(yes)	(yes)	(yes)	yes
Goods vehicles	yes	no	n/a	no	no

with five broad groups: rail and underground, parking, roads, waiting links, walk and access.

To represent the network situations at different times of day, the number of links is expanded so that there is a separate link for each time period, though the length and connectivity do not vary.

The network handles the various supply effects: capacity restraint functions modify link travel times according to traffic levels, decline in comfort on rail and underground services is represented by increasing the value of time as a function of the level of overcrowding, and the choice of parking is made between private non-residential (PNR), on-street, and public off-street parking, with costs reflecting the duration at the destination, and with search time a function of capacity utilization.

8.2.1. *Treatment of demand*

At each level in the nested incremental logit model, the choice depends on the (generalized) cost of the available options. The order of choices in the hierarchy requires that the most cost-sensitive choices are at the bottom.

At any level in the hierarchy, the measure of generalized cost needs to reflect all the choices that are implicit beneath it. This composite cost is calculated according to the well-known logsum formula (see Ben-Akiva and Lerman, 1985). Each combination of travel purpose and person type is treated separately for demand estimation purposes, and different choices are allowed, in the light of modelling complexity and empirical evidence about likely responses.

In order to deal with the potential effects of congestion charging on both mode choice and time of day switching, all home-based purposes are represented as primary destination tours (a full-scale trip-chaining model was not considered feasible). This relies on evidence that some 80% of home-based tours consist of a simple outward trip followed by a return trip home. The treatment of non-home based (NHB) movements is described below.

The demand responses allowed within the model are summarized in Table 8.2. As can be seen, all home-based purposes allow shifts between modes.

The sensitivities depend not only on journey purpose, but also on the detailed set of person types, allowing, for instance, different reactions according to income level and the level of employer-assisted motoring. Time of day choice is restricted to the HBW and HBO purposes. In the case of HBO movements allowance is also made for a change in destination and frequency. For the remaining HB purposes it was considered acceptable that in the medium term both the frequency and the destination were fixed.

NHBEB trips are treated as fixed trip matrices by mode for each time period, with changes in route as the only possible response. The logic is that most of these trips are dependent on the car, and that, like HBEB trips, they are unlikely to shift their time of travel. For NHBO trips made by car, a mechanism was devised to link the number of such trips to changes in the use of the car for home-based purposes: details are given in Bates (1994). Like NHBEB, goods vehicles have no response other than route choice. According to the route they choose, these vehicles, together with buses, which are assigned to fixed routes, will have an effect on the overall speed of traffic.

The full demand matrices for each purpose and person type are passed to the supply model (effectively 'loaded' on to the transport system), leading to further changes in generalized cost, with consequent effects on demand. On the highway side, a stochastic equilibrium assignment method is used, with different user classes having different values of time: more details are given in Williams and Bates (1993). The system iterates until an equilibrium is found, based on generalized cost.

For the most part, the model has not been directly calibrated to observed data. Instead, parameters have been selected on the basis of other modelling experience, and then the sensitivities implicit in the structure have been adjusted to reflect external expectations about elasticities. However, the time of day choice model was based on a specially commissioned project which collected stated preference data: details are given in Polak et al. (1993).

Table 8.3 summarises the overall sensitivity of the model, after all adjustments have been made. All day elasticities are given, separately by mode, for two key policy variables: petrol price, and public transport fares. These results were considered acceptably close to the available evidence for elasticities in the London context.

8.3. The testing of options

In this paper we will concentrate on so-called point-based charging systems, which have the general characteristic that the charge is levied when a specified point is passed. In all but the most complicated variants, the charge levied is independent of the charges incurred by the same vehicle at any other charging points. For a variety of technological reasons, point-based systems are easier to implement than area-based charging systems. Although they are perceived

Table 8.3. Fare and cost elasticities implied by APRIL model.

Elasticity with respect to:	Mode	HBEd	HBEB	HBW	HBO	All trips
Public transport fares	Car	+0.15	+0.07	+0.21	+0.14	+0.15
	Bus	−0.62	−0.36	−0.52	−0.48	−0.46
	Rail	−0.42	−0.16	−0.26	−1.02	−0.41
	All mechanized modes	−0.20	−0.01	−0.05	−0.14	−0.05
Petrol costs	Car	−0.33	−0.00	−0.11	−0.10	−0.12
	Bus	+0.14	+0.08	+0.17	+0.07	+0.10
	Rail	+0.13	−0.03	+0.06	+0.06	+0.05
	All mechanized modes	−0.10	−0.00	−0.01	−0.05	−0.06

as suffering from boundary problems, they can be rendered more continuous by increasing the density of charging points. In addition, they allow for variation with the direction of travel that is generally much more difficult to achieve with area-based systems. In general, the location of the points will be determined by defining appropriate cordons or screenlines through the highway network. The simplest variant, which we shall examine in some detail, is where a single cordon is drawn round the central area, and a charge is levied for crossing the cordon, in an inbound direction, at defined times.

For any defined point-based charging option, the procedure for APRIL was the same. A notional charging link was inserted in the APRIL network at each appropriate location, and vehicles traversing the link incurred the charge as an addition to their operating cost. The charge was allowed to vary by direction and time of day, and, in principle, by vehicle type, though the majority of options did not make use of this last variation. Because weekend travel was not modelled, it was assumed that charges would apply to weekdays only. This change in vehicle operating cost was then converted to generalized cost/person by dividing by the occupancy of the vehicle, which in itself varied by purpose and time of day. The resulting change in generalized cost led to changes in demand, leading in turn to a change in supply-side costs, with iterations until an equilibrium was reached.

The main approach towards the assessment was to ask: 'how would transport in London have performed had congestion charging been implemented in 1991 (the base year for the transport modelling system)?' A selected set of options was also tested in a 2011 scenario, but those results are not reported here.

8.3.1. *Environmental effects*

We concentrate on two environmental effects CO_2 and CO emissions. The former is assumed to be proportional to fuel consumption, which is the indicator actually presented here. We also present a rough estimate of accident reductions. More detailed assessments, related to local air quality and noise and severance, for example, require a model with more spatial disaggregation: such results were investigated with other models available to the study, but are not reported here.

It was considered that the most suitable source for estimating the required outputs was the report of European Commission's CORINAIR Working Group (European Commission, 1989). The group aimed to develop a baseline methodology setting out rules which could be generally applied on the basis of available statistical data in the member countries. While all vehicle types are covered, the report concentrates on emissions from petrol-driven vehicles, since overall these are the largest contributing group.

We begin with an estimate of fuel consumption. For petrol-driven cars (less than 3.5 tonnes weight) CORINAIR propose a relationship for use with engine size in the range 1.4–2.0 l at speeds up to 60 km/h of $f_{car} = 718v^{-0.694}$ (European Commission, 1989; Table VI.1.1–4), where f is fuel consumed in tonnes/million veh-km, and v is speed in km/h. The form of this function is such that fuel consumption at typical average speeds in Central London (15 km/h) is over 10 times higher than the consumption in the optimum speed range (about 60 km/h). This relationship is appropriate for 'hot running' (i.e. with the engine at normal operating temperature, assumed to imply a water temperature at or above 70°C). According to CORINAIR (European Commission, 1989; Table VI.2-2), at an ambient temperature of 10°C, fuel consumption is 35% higher (cold running). The relationship therefore needs to be adjusted to reflect the proportion of veh-km carried out under cold running conditions.

On the assumption that the average car takes about 4 km to warm up, a figure of 33% was provided by the Department of Transport for the proportion of total distance travelled relating to the first 4 km of the journey. Writing this proportion as x, the revised relationship becomes:

$$f_{car} = 718v^{-0.694}\,(1 + 0.35x)$$

For diesel cars, the relationship proposed by CORINAIR (European Commission, 1989; Table VI.1.2-1) is 57 tonnes/million veh-km, and no account is taken of either vehicle speed or cold running. For the base year for which results are reported (1991), it was assumed that only 3.5% of car-km in the London area were performed by diesel cars. Writing this proportion as z, the final relationship becomes:

$$f_{car} = 57z + (1 - z)718v^{-0.694}(1 + 0.35x)$$

This formula was multiplied by the total car-km (in millions) forecast for each link within the APRIL highway network.

For buses and commercial vehicles, simpler formulae were used. Table VI.1.5-1 of the CORINAIR report gives fuel consumption for buses as 40 l/100 km for urban conditions: assuming a fuel density of 0.825 kg/l, this translates to 330 tonnes/million veh-km. The same table, and Table VI.1.4-1 of this report give figures of 26 and 44 l/100 km for commercial vehicles of weight 3.5–16 tonnes and >16 tonnes, respectively. However, the APRIL figures for commercial vehicles also include light vans, which will have lower consumption. Based on figures for the average size of CVs in London traffic (Department of Transport, 1992), an average figure of 157.5 tonnes/million veh-km was derived. The resulting estimates of total fuel consumed are simply multiplied by a factor of 3.1 to translate them into estimates of tonnes of CO_2 produced.

To estimate CO emissions, we proceeded along similar lines. Once again, only petrol cars are treated in detail within the CORINAIR working group formulae. The corresponding relationship for use at speeds up to 60 km/h is $CO_{car} = 214v^{-0.876}$ (European Commission, 1989; Table VI.1.1-1), where CO is tonnes CO emission/million veh-km, and v is speed in km/h. In this case the relationship between emissions and speed is less steep than in the case of fuel consumption/CO_2, though still significant. For CO, however, the correction for cold running is much more important. CORINAIR's Table VI.2-2 shows that CO emissions are three times as high under cold running, at an ambient temperature of 10°C. The modified relationship thus becomes

$$CO_{car} = 214v^{-0.876}(1 + 2x)$$

where, as before, x is the proportion of total distance travelled relating to the first 4 km of the journey, assumed to be 33%.

The presence of a catalytic converter results in a marked reduction in CO emission, assumed to be 80%. The formula can therefore be further adjusted for this. If c is the proportion of petrol cars fitted with catalytic converters, assumed by the Department of Transport to be 0.5% in 1991, though rising to apply to the whole vehicle fleet by 2011, the formula can be further modified to

$$CO_{car} = 214v^{-0.876}(1 + 2x) \cdot (1 - 0.8c)$$

For diesel cars, the relationship proposed by CORINAIR (Table VI.1.2-1) is 1.0 tonnes/million veh-km, and, as with fuel consumption, no account is taken of either vehicle speed or cold running. Hence, taking the proportion of diesel cars as z, as before, the final relationship becomes:

$$CO_{car} = 1.0z + (1 - z) \cdot 214v^{-0.876}(1 + 2x) \cdot (1 - 0.8c)$$

Again, this formula was multiplied by the total car-km (in millions) forecast for each link within the APRIL highway network.

For buses and commercial vehicles, the CORINAIR report gives simpler formulae. Table VI.1.5-1 gives CO emissions for buses as 6.6 tonnes/million veh-km. The same table and Table VI.1.4-1 give figures of 7.3 and 6.0 tonnes/million veh-km for commercial vehicles of weight 3.5–16 tonnes and >16 tonnes, respectively. Making an allowance for light vans, as before, an average figure of 3.9 tonnes/million veh-km was derived.

Table 8.4. Charge levels and anticipated reductions in traffic.

	Low	Medium	High
Central	4–11%	11–18%	18–25%
Inner	3–8%	8–13%	13–18%

Once again, the resulting formula was applied to each link within the APRIL highway network.

8.4. The reference options

In this Section we report in some detail the assessment of a set of relative straightforward congestion charging alternatives. Our initial work suggested that Outer London as a whole had insufficient congestion to merit charging in the short term. We therefore concluded that we should limit our main series of tests to Central and Inner London, but with some variations in definition of area. This still left a wide range of combinations of area, time of day, and charge levels. To simplify comparisons, we specified a set of Reference Options, based on: inbound cordon charging, all day charging into Central London and as an option, charging during the peaks into Inner London. Rather than investigate continuous variations in the charge level, we specified three levels which we would expect to achieve reductions in morning peak vehicle-kilometres in Central London, and, when charged, in Inner London (Table 8.4).

It should be stressed that these levels of reduction were only chosen to give a reasonable spread for illustrative reasons and to include levels of traffic reduction which appeared likely to generate significant benefits: they have no more substantive justification. For the reference options, these charge levels were £2, £4 and £8 in Central London, all day, and £1, £2 and £4 in Inner London in the peaks.

The reference set consisted of seven tests. Four of these (charge structure A) involved charging solely to enter Central London, but applied throughout the day (07.00–19.00). Three (charge structure B) also included charging to enter Inner London in the morning and evening peaks (07.00–10.00; 16.00–19.00). For reasons of space, the results reported are confined to only four of these tests which demonstrate the range: these are indicated with an asterisk in Table 8.5.

Table 8.6 summarizes the transport effects of the four asterisked cases, showing several aggregate measures of vehicle-km and vehicle trips. For each measure, the table provides the figure in the base case, without congestion charging, in absolute units, followed by results for a set of congestion charging options, each given as a percentage change from the base. Trip-km is the number of person trip-km/day over the whole study area, in thousands, separately for car,

Table 8.5. Set of reference tests.

Charge structure	Charge level	Central London cordon charge (inbound, all day)	Inner London cordon charge (inbound, peak only)
A	low	£2*	–
	medium	£4	–
	high	£8*	–
	–	£10	–
B	low	£2*	£1
	medium	£4	£2
	high	£8*	£4

bus and rail, and also summed for all mechanized modes. Changes in the total of all mechanized trip-km reflect a shift between mechanized and slow modes, as well as four other mechanisms of lesser importance: generated/suppressed trips for certain home-based other (HBO) purposes; changes in non-home-based other (NHBO) trips consequent on changes in the choice of the car for home-based journeys; changes in the proportion of home-based journeys returning home late in the evening and changes in destination for HBO travel.

Vehicle trips is the number of trips, in thousands, made in the study area by vehicles on the highway over the whole day (06.00–20.00), separately for trips (in either direction) between each of the four areas: central (C), inner (I), outer (O) and external (X).

For the low charge with a central cordon, overall car-km would fall by only 1%. However, the high charge reduces external vehicle trips to the centre by almost a quarter, and those from the outer area by 30%. The high proportion of CV travel between the inner and central areas, which is largely undeterred by the charge, accounts for the lower effect for this movement. The low charge has somewhat less than half the effect of the high charge.

Extension of charging to Inner London in the peak periods (structure B) generally more than doubles the reduction in car trip-km. However, there is little effect on vehicle trips to and from Central London, and in some cases there is a slight (relative) increase, probably because of the reduction in congestion. Note that although travel within the central area is not charged, there is a predicted reduction which is due to the knock-on effect of reduced car availability for NHB trips. More details are given in Bates (1995).

Table 8.7 shows the effects on traffic and speeds within the three key areas. In the case of the central and inner areas, a further breakdown is given by time of day. Note that since commercial vehicle traffic does not vary appreciably according to policy (though routing variations will have some effect), the changes in car-km (not reported here) are generally greater than those seen in

Arora & Gangopadhyay – publisher?

Table 8.6. Effects of reference cases on total travel.

	Base level	Percentage change			
		Charge structure A		Charge structure B	
		Low level	High level	Low level	High level
Trip-km (millions)					
Car	125.8	−1.0	−2.4	−1.9	−5.7
Bus	17.2	2.2	5.9	3.7	11.0
Rail	43.2	0.2	0.9	1.2	4.2
Mechanized	186.2	−0.5	−0.9	−0.7	−1.8
Vehicle trips ('000s)					
C–C	253.3	−2.5	−6.8	−1.4	−4.8
C–I	401.6	−10.1	−22.9	−9.3	−22.1
C–O	211.7	−11.8	−30.9	−13.9	−30.8
C–X	98.5	−9.9	−24.9	−9.3	−20.9
I–I	1393.6	0.7	1.5	0.4	0.9
I–O	1097.3	0.8	2.1	−3.7	−12.9
I–X	238.5	0.8	2.9	−1.6	−7.1
O–O	4280.4	0.1	0.2	0.1	0.4
O–X	1265.6	−0.3	0.3	0.1	0.7

overall vehicle-km. The top part of the table presents results for vehicle-km. Note that variations in effect by time of day reflect both the possibility of time-shifting and the different sensitivities to charges and changes in conditions associated with different kinds of travel at different times of day. The bottom part of the table focuses on car speeds, but the last items give some indication of conditions on public transport, by comparing changes in total public transport demand. Speeds for commercial vehicles will be similar, while buses are modelled as travelling at 80% of the speed of cars. The effects in this table are generally the inverse of the traffic effects shown above.

At the end of the table, as an indication of the effect on public transport conditions, for the morning peak only, results are given for rail passenger-km, for rail trips destinating in the central area, and for bus passenger-km, for bus trips destinating in the central + inner areas, and the central + inner + outer areas. Demand for rail travel is greatest for travel to Central London, whereas demand for bus travel is more widespread throughout London – hence the different demand indicators adopted for the two modes.

For central charging only, the high charge reduces vehicle use in Central London by over 20%, although there is an increase of 10% in the pre-peak period. Changes elsewhere involve small reductions. The low charge achieves about 40% of the reductions obtained by the high charge. The high charge increases speeds in Central London by 30% overall, and by almost 50% in the morning peak; changes elsewhere are much smaller, but they are still positive.

Table 8.7. Impacts of reference cases on traffic by time of day.

		Base value	Percentage change			
			Charge structure A		Charge structure B	
			Low level	High level	Low level	High level
Veh-km (000s)						
Central area	overall	4064	−8.0	−21.7	−7.7	−20.2
	pre-peak	117	3.1	9.7	4.0	10.8
	am peak	843	−8.2	−23.1	−9.2	−23.0
	interpeak	1330	−10.1	−26.2	−9.0	−23.9
	pm peak	871	−8.5	−24.1	−8.7	−22.0
Inner annulus	overall	27830	−1.2	−2.9	−3.3	−9.4
	pre-peak	783	1.1	3.5	2.2	8.1
	am peak	6553	−1.0	−2.4	−5.3	−16.5
	interpeak	7886	−1.7	−3.9	−2.2	−5.6
	pm peak	6828	−1.3	−3.2	−5.6	−16.1
Outer annulus	overall	90771	−0.5	−1.0	−1.1	−3.0
Car speeds (km/h)						
Central area	overall	17.7	10.0	31.3	10.8	31.3
	am peak	14.6	10.6	47.6	12.5	49.1
	interpeak	20.0	11.7	25.9	11.6	25.0
	pm peak	16.7	9.2	32.6	9.5	31.0
Inner annulus	overall	21.1	1.9	4.2	3.5	6.8
	am peak	19.1	3.0	7.1	6.6	13.4
	interpeak	22.4	1.4	2.3	1.6	2.7
	pm peak	20.8	1.5	3.9	3.2	7.0
Outer annulus	overall	33.0	0.4	0.7	0.8	1.5
Public transport km (am peak) (000s)						
Rail	to centre	9791	1.8	4.3	2.1	3.4
Bus	to C + I	2150	7.0	13.2		
	to C + I + O	5304			5.4	16.7

Rail flows rise by around 4% and bus flows by 20%. The low charge achieves changes of about one third of those with the high charge.

Extension of charging to Inner London (structure B) has no further effect on vehicle-km in Central London, but it dramatically lowers peak and interpeak travel in the inner annulus. The high charge reduces vehicle-km in the inner annulus by 9% overall, and by 17% in the morning peak. Reductions in the outer annulus are around 3%. The low charge has around one-third the effect of the high charge in both these areas. Despite the additional reductions in flow, the increases in inner annulus speed as a result of inner area charging are relatively small. The high charge achieves an increase of speed in Inner London of 7% overall and of 13% in the morning peak. These, and the smaller increases in Outer London, are around double those achieved without inner area charging.

Table 8.8. Environmental effects on reference cases.

	Base value	Percentage change			
		Charge structure A		Charge structure B	
		Low level	High level	Low level	High level
Fuel consumption (tonnes)					
Central	542	−7.8	−20.5	−7.8	−19.5
Central + Inner	3734	−2.4	−5.9	−4.3	−10.6
CO emissions (tonnes)					
Central	67	−20.0	−46.5	−20.0	−44.3
Central + Inner	571	−5.1	−11.7	−8.5	−20.0
Accidents					
Whole area	124	−1.0	−2.6	−1.8	−4.9

8.4.1. *Environmental effects*

Table 8.8 presents the environmental consequences of the predicted changes in road traffic, concentrating on fuel consumption, CO emissions and accidents. The table gives the base figures in tonnes, over the whole day, for fuel consumption and CO emissions, separately for the two areas, and the base number of personal injury accidents/day for the whole study area. Then, for each policy, the percentage change is shown.

For central cordon charging only, the high charge reduces fuel consumption (and CO_2) in Central and Inner London by 6% and CO by 13%. Accidents within the M25 are reduced by 3%. Within Central London, fuel consumption falls by 20% and CO emissions by 45%. The low charge produces about half this effect on all indicators. For the extension to the Inner London cordon charging (structure B), the effects on overall fuel consumption, CO and CO_2 emissions and accidents are almost double those for Central London charging. Reductions of 10% in CO_2 and 20% in CO are achieved across Inner London at the high charge level.

8.4.2. *Overall commentary on reference options*

In this section the results presented for the reference options are set into context, making some use of material in the final report which has not been reproduced here.

The general effect is that car traffic with a destination in Central London (and with Inner London charging, Inner London in the peak) is discouraged, leading to reductions in car-km there. There is a slight increase in car use for trips within Central London, which are not charged, but this effect is small. The effects on vehicle-km are only about 50% of those on car-km in Central London, and 70% in Inner London, reflecting the large volumes of other traffic.

The long distance car movements divert predominantly to rail. However, the improvement in bus performance, increasing with the level of charge, means that most Inner to Central car movements which transfer change to bus, and that bus also attracts travel from rail and the slow mode. The net result is that the overall rail share does not increase much, although the effect on rail-km is greater.

Charges can be avoided altogether by shifting to the morning pre-peak period: a small amount of shifting does occur. Though the percentage increase is large, it is from a very small base. Charging limited to Central London produces only a small reduction in car traffic in Inner London, since most of the traffic there is not destined for Central London. In Inner London the percentage increases in speed are similar to the percentage reductions in vehicle-km, but rise less rapidly. In Central London, however, the percentage increases in speed are around double the percentage reductions in vehicle-km, and rise more rapidly; for the highest charges, speed increases are about 2.5 times the reduction in traffic levels. Extension of charging to the much larger inner area, even limited to the peak, substantially increases the improvements in environmental indicators for Inner London.

It is also of interest to note the effects on economic benefit, defined here, consistent with the general UK Department of Transport principles of cost-benefit analysis, as the consumer surplus measures for travellers, using the 'rule of a half' approximation, plus changes in revenue net of operating costs for 'producers'. The measure does not include environmental benefits.

Note that for a full cost-benefit analysis, we require estimates of the implementation, replacement and operating costs of the congestion charging system itself, and these will vary, of course, with the details of the technology and the scope of the scheme. Any economic assessment in this paper is primarily concerned with the variation in benefits, which are generally more related to the charges levied than the technology by which they are collected. Readers who are interested in the evaluation of different systems are directed to the full report (MVA, 1995).

It can be shown that while economic benefit initially increases with the charge level, there is a rapid reduction, for central area charging, once the economic optimum is passed. The economic optimum of around £6 is similar to those from earlier studies.

Car travellers increasingly incur disbenefits as the charge rises, since the money disbenefits rise almost as rapidly as the charge, while the time benefits rise at a decreasing marginal rate. Because freight and taxis have no mechanism to avoid the charges they are generally disadvantaged, since the time and reliability benefits are insufficient to offset the money disbenefits. The charges are predominantly borne by the high income group and employer's business trips, and their time benefits, even when valued more highly, are insufficient to compensate. For lower and middle income groups, there are small time benefits due to improved bus speeds.

8.5. Investigations of alternative charging structures

Having established the general properties of the simplest kind of charging structures (A and B), we went on to vary elements of the structure in a systematic way. As has already been noted, this paper only discusses point-based charging systems.

Key areas of investigation were extending the Inner London cordon charge to the inter-peak (10.00–16.00) (structures C, D); bi-directional charging, coupled with a range of ratios of inter-peak to peak charge (structures H–N); charging for crossing a screenline along the River Thames (structure R); experiments with multiple cordons and screenlines (structures T–X). In this Section we give a summary of the general results.

8.5.1. *Effects of extending the charge period on the Inner London cordon*

We conducted four tests with charging to enter Inner London during the inter-peak period as well as during the peak. In all cases the peak charge for the Inner London cordon was half that of the central cordon, as in the reference cases. The interpeak charge for the inner cordon was tested at a quarter and at a half of the corresponding peak charge.

The main effects of inter-peak Inner London charges are on flows in Inner London, and environmental effects generally. There is little further effect on speeds, even in Inner London. Economic benefits and revenue are increased, but the former appears to be caused primarily by the latter. There are increased disbenefits to most groups of traveller.

Overall, extension to inter-peak charging in Inner London could potentially be justified on environmental grounds, and produces some net economic benefits.

8.5.2. *Effects of charging in both directions*

We developed four alternative charging structures, H, J, K and L, in which charges are imposed in both directions. All except H vary by time of day and direction to reflect the levels of congestion experienced. There is a wide range of possible charging structures of this kind, and those tested are simply illustrative of these options. All were, however, designed to impose the same charge on a journey inbound in the morning peak, and outbound in the evening peak.

At the high charge level, the charges imposed under structures H, J, K, and L are as shown in Table 8.9. There are no charges before 07.00 or after 19.00. At the low charge level, the charges are always a quarter of the values given in Table 8.9.

Because all these charging systems apply to both directions, with mirror image charges in the evening peak, there is less benefit in transferring to the pre-peak, in contrast to uni-directional charging, where it is possible thereby

Table 8.9. Bi-directional charging structures for Central London cordon, high charge level.

Charge structure	07.00–10.00	10.00–11.00	11.00–15.00	15.00–16.00	16.00–19.00
H					
Inbound	£4.00	£4.00	£4.00	£4.00	£4.00
Outbound	£4.00	£4.00	£4.00	£4.00	£4.00
J					
Inbound	£4.00	£3.00	£2.00	£1.00	–
Outbound	–	£1.00	£2.00	£3.00	£4.00
K					
Inbound	£4.00	£2.67	£2.67	£2.67	£1.33
Outbound	£1.33	£2.67	£2.67	£2.67	£4.00
L					
Inbound	£4.00	£2.00	£1.00	£1.00	–
Outbound	–	£1.00	£1.00	£2.00	£4.00

to avoid the charge entirely: with bi-directional charging, the working day would have to be extended at both ends to avoid the charge. Consequently, with bi-directional charging, similar results are obtained for the morning and evening peaks.

Uniform bi-directional charging, limited to Central London (structure H), performs very similarly to uni-directional charging. Traffic reductions are slightly greater, but overall speed, environmental and safety effects slightly smaller. Economic benefits are between 5% and 15% higher, and traveller benefits are increased. However, the additional complexity of bi-directional charging is probably not justified if charges are kept uniform.

Variable bi-directional charging, limited to Central London, performs very similarly to uniform bi-directional charging, and hence to uni-directional charging. Structures J and K generally have a similar effect, with smaller reductions than uni-directional charging in traffic, environmental and safety indicators, and smaller increases in speed. Economic benefits are lower at the low charge, but higher at the high charge. The most important difference is in traveller benefits, which are positive at the low charge, and very much less negative at the high charge. Structure L performs markedly less well on most indicators, since it imposes lower inter-peak charges, but achieves positive traveller benefits at both charge levels.

The variable bi-directional charging structures were extended to the Inner London cordon (structures M and N), again on the assumptions of peak-only charges at half the corresponding central cordon charge. This produces slightly greater reductions in car use, fuel consumption, and environmental effect than the reference option, and slightly greater improvements in speeds outside Central London. At high charge levels, it produces substantially greater economic benefits, and lower disbenefits to travellers. This can be illustrated

Table 8.10. Selected results for alternative charging structures (high charge level).

	Charge structure				
	Inbound only		Bi-directional		
	B	R	M	T	W
Veh-km (% change)					
Central area	−20.2	−19.5	−16.9	−17.1	−16.7
Inner annulus	−9.4	−11.6	−10.1	−9.2	−10.8
CO emissions (% change)					
Central	−44.3	−42.8	−40.1	−40.4	−39.5
Central + Inner	−20.0	−23.1	−20.6	−20.2	−22.7
Economic benefit (£ million/annum)					
Travellers	−352.4	−398.0	−173.3	−141.2	−151.0
Total[a]	277.6	310.1	377.7	393.0	445.6

[a]Travellers benefit + accidents savings + gross revenue: does not take account of implementation and collection costs.

B, 2 inbound cordons: uniform charging; M, 2 bi-directional cordons: variable charging; R, 2 inbound cordons with screenline: uniform charging; T, 3 bi-directional cordons: variable charging; W, 3 bi-directional cordons with screenlines: variable charging.

by selected results for structure M, which corresponds with structure J, extended to an Inner London cordon with half the Central London charge: the results are shown in Table 8.10.

Overall, directional charging represents a small improvement over uni-directional charging, but significantly reduces the disbenefits to travellers. Indeed, when limited to low charges in Central London, it achieves benefits for virtually all groups of travellers and, with structure L, achieves this at high charge levels also.

8.5.3. *Effects of adding the Thames screenline*

The time-variant charging options just discussed bear out the general principle that traveller benefits are improved if the charge is better targeted at congestion. However, none of the structures tested charge travel that is wholly within a given cordon, in spite of the fact that such travel contributes to congestion.

We addressed this problem by considering the addition of a Thames screen-line to the Central and Inner London cordons with charging to cross the river in either direction within Inner (but not Central) London (structure R). The tests were based on uni-directional, inbound charging for the cordons.

The Thames screenline is primarily effective in achieving further reductions in congestion in Inner London. Environmental and economic benefits and revenue are all around 10–15% higher, and there is little change in the distribution of benefits. This does appear to be a worthwhile improvement to basic uni-directional cordon charging.

8.5.4. *Effects of multiple cordons and screenlines*

Since the Thames screenline demonstrated the additional benefits to be gained by charging certain orbital movements within Inner London, we extended this concept further in a series of tests using three cordons and, in some cases, a series of screenlines.

The three cordons which we used were (a) the standard cordon around Central London, (b) an inner cordon inside the South Circular Road and its mirror image north of the River, and (c) an outer cordon inside the North Circular Road and its mirror image south of the River. Note that the area known as Inner London is now divided, its southern portion lying within the inner cordon, but its northern portion lying between the inner and outer cordons. The screenlines were designed to divide each annulus into four, using the River Thames, two additional screenlines north of the River in the inner annulus, and one north and one south in the outer annulus. This is illustrated in Figure 8.2. The small dashed sections in the south-west and south-east of the inner area indicate those sections of the North/South Circular road that no longer form a charging boundary.

We tested three different sets of ratios among the charges applying to the three cordons. Each used variable bi-directional charging. The first had equal charges on the inner and outer cordons, each one quarter of that on the central cordon (structure T). The second had charges on the inner and outer cordons three-eighths and one-eighth of the central cordon charge (structure U). The third had no charge on the outer cordon, and charged half the central cordon charge on the inner cordon (structure V). All three were designed so that the total cost for a tour to Central London in the morning peak and returning in the evening peak would be the same as for structure M.

The structures with three cordons generally perform less well than those with two bi-directional cordons, with the exception that Inner London speeds are higher, economic benefits (for structure T only) are higher and traveller benefits are greater. Of the three structures, structure T generally performs best.

However, when screenlines are added, the performance generally is much improved, to a level above that for the bi-directional cordon. This is because orbital movements, as well as radial ones, are affected. Structure W (with equal charges on both inner and outer cordons) performs best. It achieves improvements over the bi-directional cordon of 5% reduction in car trips, 10% reduction in car-km in Inner London, 10% improvement in environmental and safety indicators and in revenue, 15% economic benefit and 40% in speed.

The effects on car use in Central London are very similar with and without the screenlines. However, the screenlines achieve a 10% greater reduction in

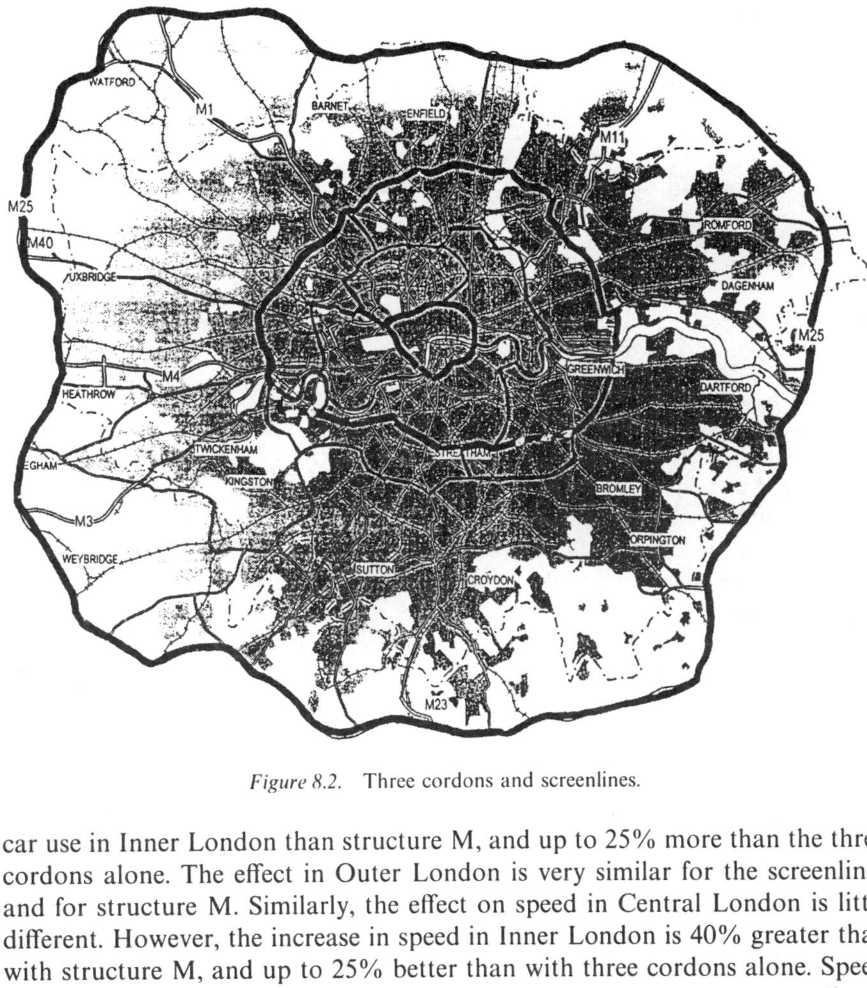

Figure 8.2. Three cordons and screenlines.

car use in Inner London than structure M, and up to 25% more than the three cordons alone. The effect in Outer London is very similar for the screenlines and for structure M. Similarly, the effect on speed in Central London is little different. However, the increase in speed in Inner London is 40% greater than with structure M, and up to 25% better than with three cordons alone. Speed improvements in Outer London are again slightly better with the screenlines.

Although structure W has virtually identical effects to structure M on emissions in Central London, the effects throughout Central and Inner London, and on London-wide accidents, are around 10% higher than for structure M.

In both cases the economic benefits at the high charge level are just over 15% greater than for structure M, and 15% greater than their equivalent three cordon tests. The same is true at the medium charge (for structure W) and in this case the total traveller disbenefits are virtually zero. At low charge levels the differences are less marked. Structure W, at the high charge level, produces the highest economic benefit of any of the tests conducted.

Table 8.10 summarizes the main results of these comparisons. In order to assist comprehension of the different structures, particularly regarding the economic results, it is helpful to set out the charges for a round trip beginning

Table 8.11. Representative (high level) daily charges (£) for alternative structures (peak/off-peak).

	Charge structure				
	Inbound only		Bi-directional		
	B	R	M	T	W
O–C	12/8	12/8	12/6	12/6	12/6
O–I	4/0	4/0	4/2	4/2	4/2
I–C	8/8	8/8	8/4	8/4	8/4
I–I	0/0	4/0	0/0	2/1	3/2
I–O	4/0	4/0	0/2	0/2	0/2
C–I	8/8	8/8	0/4	0/4	0/4
C–O	12/8	12/8	0/6	0/6	0/6

in area X and with destination in area Y, where X and Y can be Central, Inner or Outer London. Table 8.11 gives charges under each structure for peak journeys and an illustrative average off-peak journey. For all structures, the peak charges for journeys O–C and I–C are £12 and £8 respectively: however, other charges vary considerably. In particular, the bi-directional charging structures illustrated assume that peak journeys made in the direction opposite to that of the main flow incur no charge, while structures R, T and W intercept some journeys within the inner annulus.

8.6. Overall assessment

In this paper we have summarized the results of an extensive programme of tests, and provided brief commentaries on them. We now discuss the main messages from those tests.

8.6.1 *Reference charging structures*

The immediate effect of charging in Central London (structure A) is to reduce journeys by car to Central London. Most of these car users switch to bus, but there are also some transfers to rail, to travel before the peak, and to journeys to Inner and Outer London. There is also a reduction in car and commercial vehicle traffic through Central London, most of which is diverted to Inner London. At the highest charge tested (£10) car trips to Central London from Inner London during the peak fall by over 60%, from Outer London by almost 50% and from outside London by over 30%. Car traffic in Central London all day falls by over 40%, and all traffic by almost 25%. However, around two-thirds of these effects are achieved by a charge of only £4, suggesting diminishing returns from higher charges.

The reduction in traffic in Central London leads to increased speeds, which rise, on average over the day, by almost 35%. The traffic reductions also lead to improvements in the environment and safety. With a £10 charge London-wide CO emissions fall by over 5%, and fuel consumption, CO_2 and accidents by around 3%. In Central and Inner London the reductions are 13% and 6% respectively. Once again, however, around two-thirds of those effects are achieved by a charge of only £4.

Total net economic benefits rise as charges increase to a charge of around £6. Beyond that, the losses in benefits from further changes in travel exceed the benefits from further reductions in congestion. Virtually all of the achievable economic benefit has been obtained by a charge of £4/crossing. However, the diminishing returns are less strong in the case of the environmental indicators.

Extending inbound cordon charging (at half the Central London charge) to Inner London in the peaks (structure B) accentuates the effects of Central London charging. There is a further reduction in car use, most of which switches to bus and, to a lesser extent, rail. The effects on car traffic and overall traffic in Central London are similar to those with structure A, although there is a minor transfer back to car as a result of reduced congestion.

The greater reduction in Inner London traffic does not substantially improve Inner London speeds, which are only 7% faster at the highest charge level. The intermediate charge achieves three-quarters of this effect. The environmental and safety effects are, however, almost double those obtained from Central London charging, because the area affected is so much larger. The highest charge reduces London-wide CO emissions by 9%, and fuel consumption by 5%. In Central and Inner London the reductions are 20% and 11% respectively. Once again, the intermediate charge achieves around two-thirds of these effects.

8.6.2. *Alternative charging structures*

While the concept of cordon charging is simple, its application offers a wide range of options. Here we summarize the tests by focusing on the main differences between them in terms of economic benefits, CO reduction (which proves to be a reliable proxy for environmental effects generally) and traveller benefits.

Figure 8.3 shows the results of all the tests (including both the reference and the alternative structures) for both percentage CO reduction in Central and Inner London and total economic benefit. In general there is a roughly linear relationship, indicating that increasing environmental benefit can normally be obtained concomitantly with increasing economic benefit, to some extent by increasing the charge level, and to some extent by widening the scope of the charging scheme, either in area terms or by time of day.

Interestingly, however, the figure reveals a few identified outliers in which greater environmental improvements can be obtained for the same economic benefit. These are, in all cases, the highest charge levels for the inbound charge options. Charges have been raised to the level at which little more economic benefit can be gained, while the further reduction in car use continues to

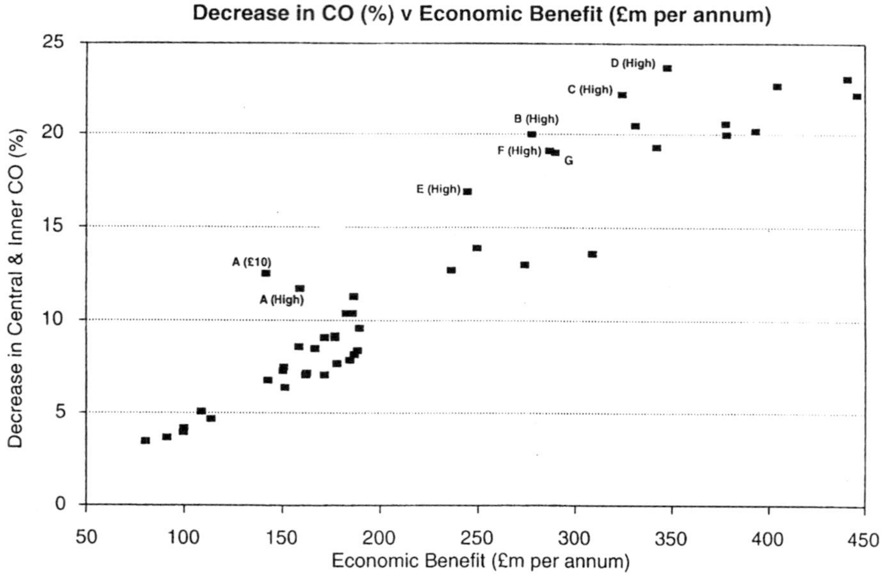

Figure 8.3. 1991 congestion charging tests – decrease in CO (%) v. economic benefit (£m/annum).

increase environmental benefits. Indeed, the figure makes one particular example clear: for charge structure A, increasing the inbound charge from £8 to £10 reduces overall economic benefit, but continues to make environmental improvements. There is thus, at these high charges, a potential conflict between the criteria for assessment.

We conclude from this analysis that it is worthwhile to extend charging to Inner London in both the peak and inter-peak periods; increasing charge levels up to the economic optimum for the charging option will continue to generate worthwhile increases in environmental benefits; bi-directional charging, particularly when extended to Inner London, is markedly preferable to inbound charging in generating higher benefits and smaller traveller disbenefits; however, performance is quite sensitive to the distribution of charges by direction and time of day; while a third cordon adds little on its own, a third cordon with screenlines in Inner London adds significantly to economic and environmental benefits and reduces traveller disbenefits.

The best performing option which we tested involves three cordons and four Inner London screenlines, with charges on each Inner London cordon one quarter of those on the Central London cordon (structure W). At high charge levels (of £4 in the peak directions in the peaks) this increases speeds by 26% in Central London and 10% in Inner London, reduces London-wide CO by 10%, and fuel consumption by 5%, with roughly double these effects in Central and Inner London. It achieves total economic benefits of £450 million/annum with traveller disbenefits of £150 million p.a. (and significant benefits for medium and low income travellers) and generates revenue of £740 million p.a. A charge of half this level achieves 70% of this economic benefit, and 55% to 60% of the speed and environmental benefits and revenue. It generates virtually no traveller disbenefits, net, although there would be losers as well as gainers.

Note

1. Independent Consultant, Oxford, UK. The author had the role of Technical Adviser for Modelling as part of The MVA Consultancy team, who were the Programme Consultants for the UK Department of Transport's London Congestion Charging Research Programme. The paper in this volume is largely an edited version of the text of Chapters 4, 5, 8 and 10 of the Final Report (MVA, 1995), to which many other members of the team made substantial contributions. The paper is published with the permission of the Department of Transport and the Controller of Her Majesty's Stationery Office. The views expressed in it are not necessarily those of the Department of Transport or of any other government department.

References

Bates, J.J., 1994, Modelling the response of non-home-based trips to congestion charges, PTRC European Transport Forum, Transportation Planning Methods Seminar, Warwick

Bates, J.J., M. Brewer, P. Hanson, D. McDonald and D.C. Simmonds, 1991, Building a Strategic Model for Edinburgh, PTRC Summer Annual Meeting, Transportation Planning Methods Seminar.

Bates, J.J., I.N. Williams, D. Coombe and J. Leather, 1996, The London congestion charging research programme: 4. The Transport Models, traffic engineering & control, 37, 334–339.

Ben-Akiva, M.E. and S.R. Lerman, 1985, Discrete Choice Analysis: Theory and Application to Travel Demand, Cambridge: MIT Press.

Department of Transport, 1992. London Traffic Monitoring Report: 1992, London: HMSO.

European Commission, 1989, European Commission CORINAIR Working Group, H.S. Eggleston et al., Emission Factors for Calculating 1985 Emissions from Road Traffic.

MVA Consultancy, 1995, The London Congestion Charging Research Programme – Final Report, prepared for the Government Office for London, London: HMSO.

Polak, J.W., P.C. Vythoulkas, P. Jones, R. Sheldon and D. Wofinden, 1993, Travellers' Choice of Time of Travel under Road Pricing, PTRC Summer Annual Meeting, Seminar D, Transportation Planning Methods.

CHAPTER 9

Optimal Speed Limits for Various Types of Roads: A Social Cost–Benefit Analysis for the Netherlands[1]

Piet Rietveld, Arjan van Binsbergen, Theo Schoemaker and Paul Peeters

9.1. Introduction

Transport of passengers and freight is dominated by road transport which appears to have a range of advantages in terms of the quality of the services rendered that make it more attractive than other transport modes. In addition to the benefits related to road transport, there are also costs, partly in terms of private costs (costs of fuel, etc.) and partly in terms of external costs (emissions, noise, accidents). These external costs are sometimes assumed to be proportional to the total number of vehicle miles travelled, but there are several reasons why such a relationship may be questioned.

A first point concerns changes in the technology of vehicles. Technological progress has been quite successful in reducing the external effects of road transport per vehicle mile travelled. For example the introduction of catalysts in passenger cars has led to a substantial reduction in NO_x emission. A related development with mixed effects is that cars tend to become bigger and heavier, with the effect that they may become safer (for the driver), but that energy consumption will increase.

A second issue relates to changes in the road system. As the proportion of express ways in the road system increases, there will be a shift from roads with relatively high accident rates to roads with high emissions and energy consumption. Similarly, metropolitan ring roads may help in diverting traffic away from highly populated zones to less vulnerable areas leading to less noise nuisance even when the total number of vehicle kilometres driven remains the same.

Third, special infrastructure measures may reduce external effects of transport. An example is the implementation of preventive measures such as noise shields in densely populated areas.

A final issue is that the external effects clearly depend on driver behaviour. Higher speeds usually lead to higher energy consumption, more noise annoyance, higher emissions, more (severe) accidents. It is to this issue of speeds that

R. Roson and K.A. Small (eds.), Environment and Transport in Economic Modelling, 206–225.

the present study is addressed. When the pattern of speeds on roads changes the total volume of external effects also changes, even when the total number of vehicle miles travelled remains the same. Speed is not the only relevant factor here: several other behavioural components are also relevant in this context, including acceleration behaviour, maintenance of cars, the use of special equipment (like AC) in cars, etc. However, in the present paper we will focus on speed, because this is the behavioural component about which our knowledge is best developed. In addition, it appears that the effects of speed on external effects are substantial.

An important element of discussions about appropriate speed is that we should distinguish between different types of roads. The design and the functions of these roads are so different that for each type of road different optimal speeds can be expected. It is therefore important to avoid a focus of the discussion about speed limits on express ways only. When traffic accidents are considered, it becomes especially clear that much greater gains can be obtained by reducing speeds on urban roads than on express ways, which are already very safe. If one focuses on energy consumption and emissions, however, speeds on express ways are very important.

In the present paper we report a recent study carried out in the Netherlands into optimal speed limits of passenger cars. For a more complete presentation of the results of this study refer to Peeters and Van Asseldonk (1996).

9.2. Concepts of speed

Several types of speeds can be distinguished. Roads usually have a legal maximum speed, which is specific for a type of road. It may be uniform among various types of vehicles, but not necessarily so: trucks may have lower speed limits than passenger cars. Usually, no differentiation is made across types of drivers (the Dutch queen is not allowed to drive faster than her subjects; young, inexperienced drivers are allowed to drive at the same speeds as every other person).

Vehicles have a highest possible speed. Passenger cars can usually drive much faster than the maximum speed at express ways. In some countries, automatic speed control equipment has been imposed by the government to remove the gap between the legal maximum speed and the maximum speed of the vehicle. This holds true for example for trucks within the EU. It should be noted, however, that this is only helpful when the vehicle is used at roads of the highest speed category; it is at best only partially effective in reducing the gap between the legal maximum speed on other roads and the highest possible speed of the vehicle.[1]

Drivers have an individually desired speed which depends among others on individual features, car features and the type of road. The maximum possible speed of a car imposes an upper limit on the individual desired speed. A conflict occurs when the driver's preferred speed exceeds the legal speed limit. The solution of this conflict is usually sought into the direction of a system of police surveillance and fines. There is also a socially optimal speed: this is usually lower than the individually desired speed and depends on factors underlying the individually desired speed, but also takes into account external effects of transport. It is clear that the socially optimal speed in principle varies among drivers, car features and road types. It is to this socially optimum speed that the present study is addressed.

In the case of congestion, drivers are confronted with a highest possible speed on certain roads. These highest possible speeds are usually uniform across all drivers on a road at a certain time. An exception is provided when drivers have the choice to take car pool lanes (or similar lanes for specific types of users). These special lanes may be free for specific user types, but in some cases drivers may also have to pay for them. Highest possible speeds on certain roads are highly time and location dependent.

Actual average speed should also be mentioned. The average speed depends in a complex manner on the various limits implied by the above mentioned speeds. It is especially important to realize that trips are not made at a constant speed, but that speeds change frequently. This is not only caused by the fact that a route usually consists of a chain of road segments of different types, but also since during a trip many stops have to be made at crossings. Thus the actual average speed will always be (much) lower than the individual desired speed, even if the other limitations on speed would not be effective.

In addition to these maximum speeds, minimum speeds are also effective, such as the minimum speed for cars at express ways. Minimum speed limits help to reduce speed variance, which is one of the factors influencing road safety.

The basic question addressed in this paper is 'what is the socially optimal speed per road type assuming an average driver using an average passenger car?'. We will focus on passenger transport and not discuss similar questions for freight transport, including the interactions between the two themes. To address the above question we will use a social cost and benefit type of approach.

9.3. Impacts of speed limits

A simple model of the impacts of speed limits for a certain road type is shown in Figure 9.1. We assume that the number of vehicle miles on this road type is constant. For each road type, a driver has his own preferred speed, which depends on many factors (for example personal features and features of the vehicle), including the maximum speed for that road (see also Rienstra and Rietveld, 1996). The driver is assumed to drive according to his own preferred speed, but he/she has to stop from time to time, depending on the road type.[2] This leads to an average speed that is (much) lower than the preferred speed.

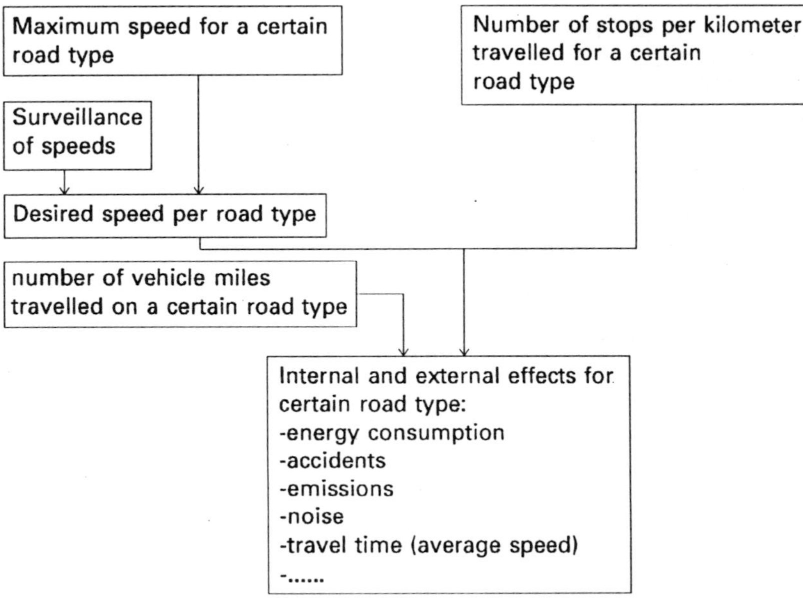

Figure 9.1. Impacts of speed limits for a certain road type (vehicle miles assumed constant).

Given the varying speeds of the drivers on the road various effects can be computed (energy, emissions, accidents, noise, and travel time). A more refined model is shown in Figure 9.2, which assumes that drivers react to the speed limits in two ways: they may choose other routes when maximum speeds on particular roads are changing, or they may decide to change also other elements of their travel behaviour (mode choice, destination choice), leading to other numbers of vehicle kilometres. Figure 9.2 provides a number of important elements when one wants to analyse the impacts of changes of speed limits in the situation of elastic transport demand.

9.4. A cost–benefit analysis approach to speed limits

We will use social cost–benefit analysis to determine optimal values for the maximum speed per road type (for an early contribution to this theme refer to ECMT, 1978). The socially optimum level of maximum speed depends on the impact of speed limits on the effects displayed in Figures 9.1 and 9.2, and on the economic values to be attached to these effects. Thus for a social cost benefit analysis one needs four types of information:

- effects of maximum speeds on actual speeds;
- effects of actual speeds on variables such as emissions, energy consumption, accidents, etc.;

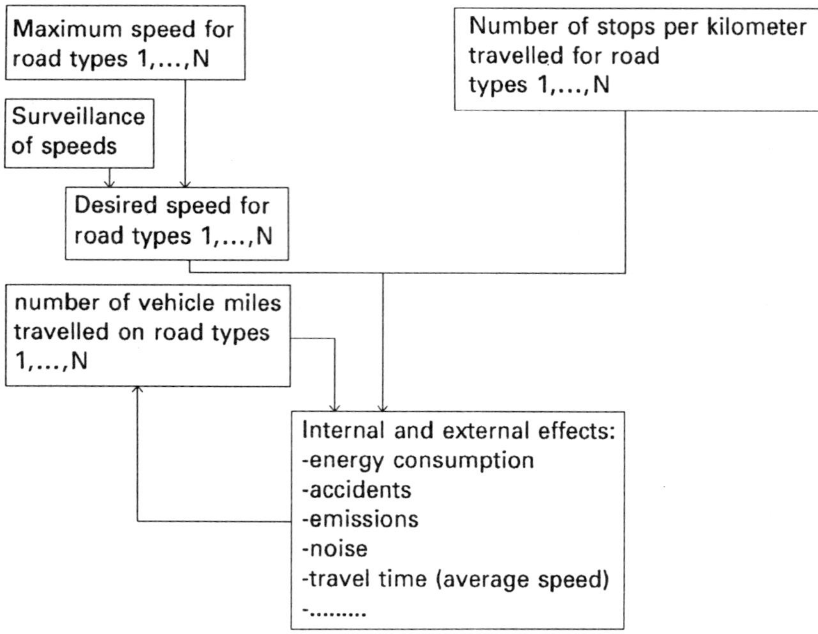

Figure 9.2. Impacts of speed limits for various road types (flexible distribution of vehicle kilometres across road types).

- economic values of these external and internal effects;
- effects of actual speeds on travel behaviour (route choice, mode choice, destination choice).

In our modelling experiments we did not explicitly model the relationship between maximum speeds and desired speeds of drivers. It is clear that such a relationship depends on many factors such as the features of roads, properties of the car, personal features, police surveillance, level of fines, etc.

With the current speed limits we carried out computations for both the actual speeds and the formal maximum speeds. References on this subject can be found in Jorgenson and Polak (1993), Rouwendal (1996) and Rienstra and Rietveld (1996). For other speed regimes we ignored the possibility of speed transgressions which means that we most probably overestimate the size of the effects, unless new instruments also introduced to force compliance of drivers.

The effects on emissions and energy depend mainly on technical features of the cars. The shapes of the functions relating speeds and emissions are shown in Figures 9.3 and 9.4. For details refer to Klein (1993) and TUD (1995). The curves for energy consumption and CO_2 emissions are assumed to be identical except for a proportionality factor.

For traffic safety a distinction is made between fatal accidents and other

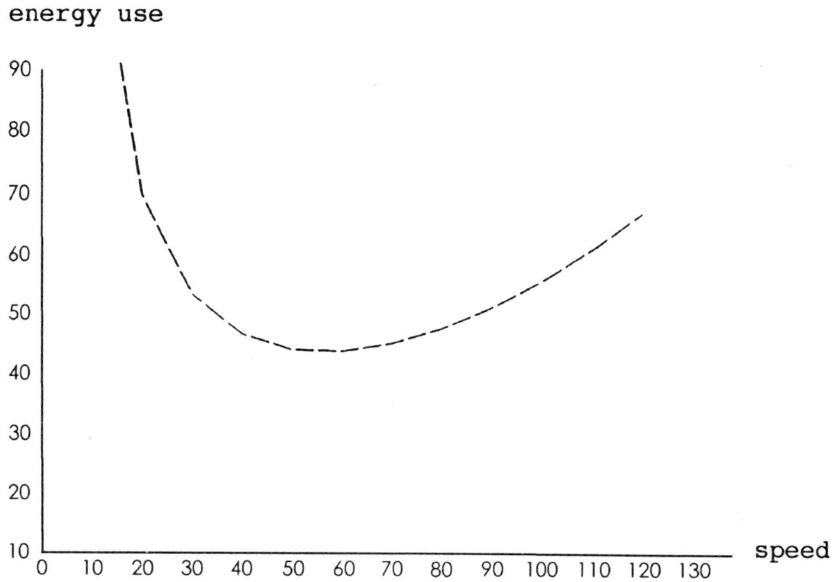

Figure 9.3. The relationship between vehicle speed (km/h) and energy use (g/km).

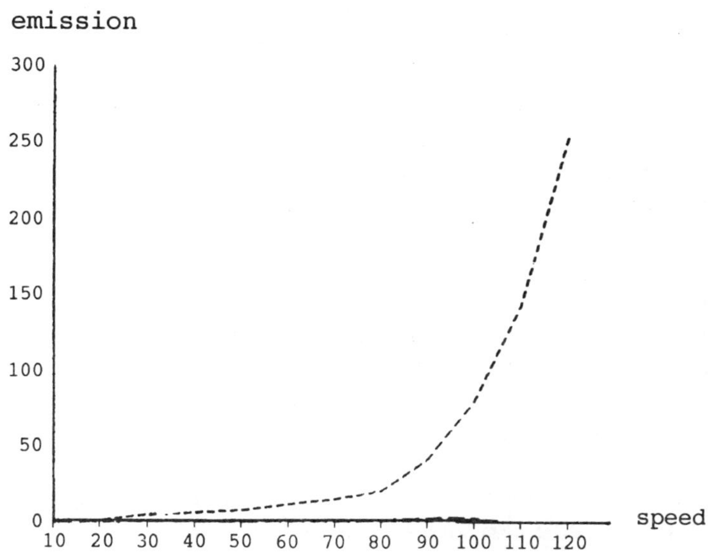

Figure 9.4. The relationship between vehicle speed (km/h) and NO_x emissions (g/km).

serious accidents leading to hospital admission. We use a simple model based on Koornstra (1993), where the number of serious accidents is assumed to be proportional to the number of conflicts (which is proportional to both the

Table 9.1. Valuation of various effects of transport.

	Unit of measurement	Minimum value	Mid-value	Maximum-value
CO_2 emission	dfl/kg	0.05	0.11	0.50
NO_x emission	dfl/kg	1.50	9.40	12.00
Serious accidents	1000*dfl/ injured person	50	137	300
Fatal accidents	1000*dfl/ fatality	400	1311	1500
Energy use	dfl/l	0.50	0.50	0.50
Travel time	dfl/h travelled	11.10	11.10	11.10

Source: Bleijenberg et al., 1994; HCG, 1990a.

differences in speeds and speed itself). In addition, the number of fatal accidents is again assumed to be proportional to both speeds and speed differences.[3] When speed differences, measured as the standard deviation of speeds are assumed to be proportional to the average speed on a certain road type, we arrive at the result that serious accidents (leading to hospital admission) are proportional to the second power of average speed, and fatal accidents are proportional to the fourth power of average speed.

Thus for each relevant effect we arrive at curves indicating the relationship between speed and the level of the effect similar to those presented in Figures 9.3 and 9.4. In most of these curves the size of the effect will increase with speed. An exception is of course travel time since travel time = distance travelled/speed, so that as long as distance travelled remains constant, travel time is proportional to the inverse of speed. The various effects can be classified as internal to the driver (travel time), external to the driver (pollution) and mixed (traffic safety).

A rather difficult topic concerns the economic valuation of the various effects. Reviews of the various methods to monetarize the effects can be found among others in Pearce and Markandya (1989). Results of these methods have been published among others in Quinet (1989), Kageson (1993), Verhoef (1994) and Bleijenberg et al. (1994). In the present study we will use the results of Bleijenberg et al.[4] The ranges of values resulting from this study can be found in Table 9.1. This table also shows the price of petrol for passenger cars. Note that this is the price of producing petrol without including the various taxes. This is the appropriate price to be used in cost–benefit analyses. The actual price including taxes paid by the drivers is almost four times the price reported here. In the last row of the table we present the value of traveltime per hour saved. These results are based on an extensive travel time study carried out by HCG (1990). The figure given here is an average based on separate values for

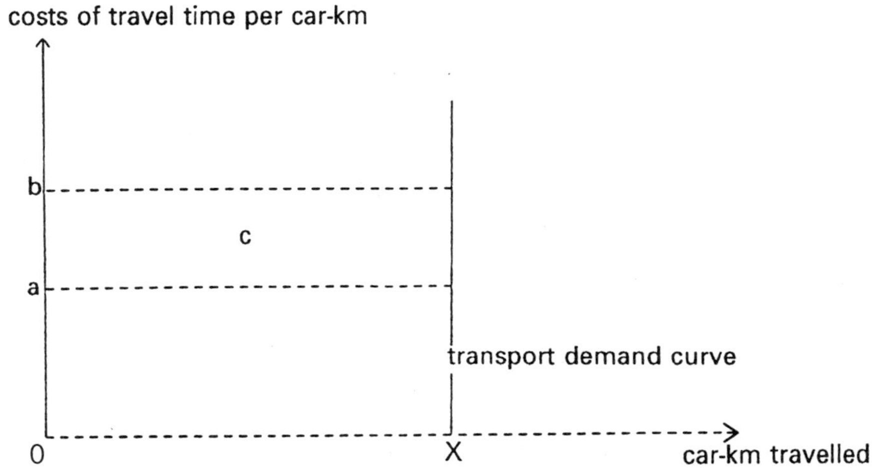

Figure 9.5. Costs of an increase in travel time with inelastic demand for transport.

business traffic, commuting and other trip purposes. For business traffic the value is about twice the average value given here.[5]

The application of social cost–benefit analysis is straightforward for the case shown in Figure 9.1 where a partial analysis for certain road types is given and where the total number of kilometres travelled is assumed to be constant. The treatment of travel time losses in this case of inelastic demand for car-km is shown in Figure 9.5. An increase in travel time/km travelled from a to b leads to an increase of travel time costs equal to C. Thus, the change in welfare due to travel time losses is simply proportional to travel time itself, which in its turn is (by definition) inversely proportional to speed as we noted above.

Application of social cost–benefit effects to determine optimal speeds require an economic valuation of the various effects. This leads to a weighted summation of the various speed–effect curves. Figure 9.6 gives a simple illustration of this for the case that we would only consider two effects (for example internal costs: travel time and external costs: NO_x emissions). The location of the socially optimal speed depends of course on the weights attached to the internal and external costs. In addition it would depend on the type of road because the speed–effect curves for safety (and noise) strongly vary across road types; in addition the number of stops per km driven also varies largely among road types.

The approach presented above is sufficient to carry out a cost–benefit analysis for the change in maximum speeds at a certain road, as indicated in Figure 9.1. However, this would only lead to a partial analysis since maximum speed limits may lead to various behavioural responses. In order to carry out such an analysis one would ideally use a full network model with a classification of all relevant road types according to speed limit. This would enable one to

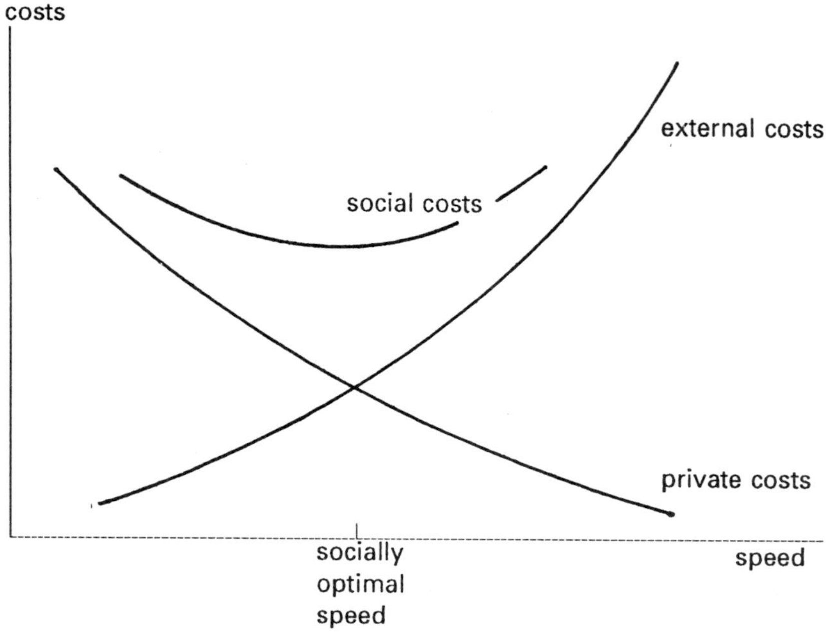

Figure 9.6. The determination of the socially optimal speed by adding private and external costs of transport.

carry out an analysis of behavioural responses in terms of route choice, etc. In the present context such a model was not available to us. Therefore a less refined approach was used by employing some key parameters in terms of general travel time elasticities of car use derived from the Dutch national transport model (the LMS model, Ministry of Transport, 1990). These elasticities indicate that indeed a certain reduction takes place in the total number of kilometres travelled by car; this is the combined result of shifts in destinations and shifts in transport modes. Thus in this first stage a change in car-km travelled is found while ignoring changes in route choice. Since we need results on kilometres travelled per road type the total change in kilometres travelled has to be allocated according to the road types. In the first stage we assume an equal proportional reduction for all road types.

In the second stage, changes in route choice are taken into account by assuming that a reduction of speed on a certain type of road makes this road type less attractive, so that car drivers will partly shift to other road types. We did this in an admittedly simple way by assuming that a reduction of speed on a road type by 1% leads to a reduction of car use on that type of road of 0.5%, and to an increase on other roads which can serve as substitutes keeping the total number of kilometres travelled (determined in stage 1) constant. The

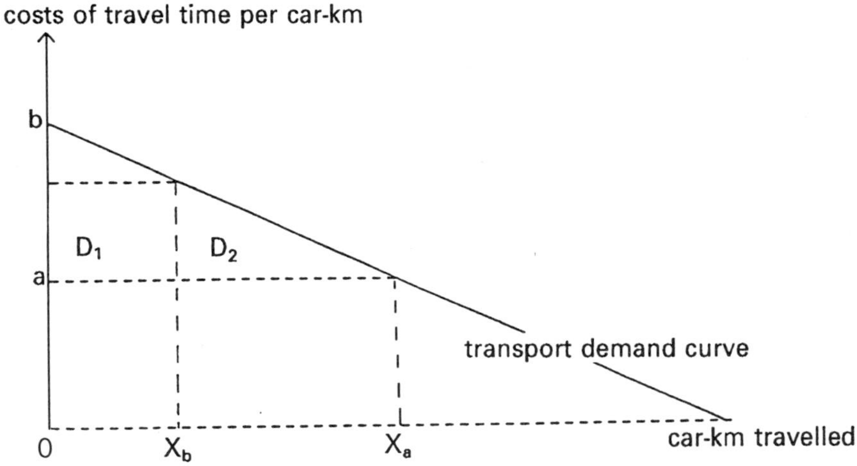

Figure 9.7. Costs of an increase in travel time with elastic demand for transport.

increase of kilometres travelled on the substitute roads is assumed to be proportional to the present shares of these road types in total traffic.

When the total number of kilometres travelled is allowed to change, we take into account that car users trade-off the utility of making their usual trips against the utility of changing mode or destination. This has welfare implications. Figure 9.7 shows that when the demand for transport is elastic in terms of speeds, total traveltime is equal to $a \cdot X_a$ before the speed reduction, and $b \cdot X_b$ after the reduction. When the travel time elasticity of demand equals -1 total travel time would remain constant. However, for a cost–benefit analysis of speed reductions we should not compute the change in monetarized total travel time, but the change in consumer surplus. This change in consumer surplus consists of two parts: D_1, the welfare loss related to trips which remain unaltered due to the speed change (comparable to the area C in Figure 9.5), and D_2, the welfare loss related to those trips where the driver found an alternative destination or used an alternative mode.

When we compare Figures 9.5 and 9.7 an important conclusion is that the time related costs of speed reductions are smaller when demand is elastic. At the same time the benefits are also more favourable (total emissions etc. would decrease when car-km decrease). Thus the outcome of a cost benefit analysis of speed reductions will be more favourable when transport demand is elastic.

A related type of effect concerns the increase in fuel efficiency of cars which is usually observed when speeds are decreased. This would lead to smaller variable fuel costs, which would in turn lead to an increase in transport demand. This effect can (partly) off-set the above favourable effect due to time-elastic travel demand.

Table 9.2. Classification of road types in the Netherlands.

Road type	Maximum speed (km/h)	Actual average speed (km/h)	Length (km); % shares in parentheses	Vehicle-km/year (% shares in parentheses)
Express ways	120	113	1784 (1.7)	23122 (25.1)
Express ways	100	105	381 (0.4)	10035 (10.9)
Highways	100	92	1154 (1.1)	4143 (4.5)
Highways	80	80	7776 (7.4)	17367 (18.9)
Other roads outside urban areas	80	64	44991 (42.9)	14212 (15.4)
Major urban roads	50	38	11331 (10.8)	18674 (20.3)
Other urban roads	50	27	37459 (35.7)	4590 (5.0)
Residential area (low speed)	30	15	2140 (2.0)	78 (0.0)
All roads		63	104830 (100.0)	92220 (100.0)

[a]Based on an occupancy rate of 1.3 persons per car.

9.5. Empirical results

9.5.1. Existing situation

The first step in the empirical analysis concerns a classification of road types. In our study we use eight types of roads (see Table 9.2).

The length of the express ways is about 2% of the total length of the Dutch road network. However, in terms of the total number of kilometres travelled its share is about 36%. The average speed is lower than the maximum speed on all road types except for the express ways with a speed limit of 100 km/h. This does not mean to say that the maximum speed is always obeyed on the other roads. Not only may there be substantial differences between the desired speeds of individual drivers, but (and this is especially relevant for the urban roads) the number of stops per kilometre is substantial, leading to a low average speed even though the desired speed of drivers may be substantially above the average speed.

Table 9.3 gives the emissions of passenger car transport in the base year on eight road types. The data in this table originate from different sources: data on accidents are based on police reports and hence directly observed. However, the reports are not always specific enough to know on what type of road the accidents take place. Hence certain rules of thumb had to be used to allocate accidents to roads. The data on travel time are based on combinations of speed observations and traffic volumes per road type. Here too, rule of thumb sometimes had to be used to make the information complete. Data on CO_2, NO_x and energy use are based on technology curves as presented in Figures 9.3 and 9.4, multiplied by numbers of vehicle-km.

When we compare the shares of the various road types in the total volume

Table 9.3. Effects of travelling on alternative road types; base situation 1992 (% share in parentheses).

Road type	Maximum speed (km/h)	CO_2 emissions (millions of tons)	NO_x emissions (thousands of tons)	Number of seriously injured persons	Number of persons killed in transport	Energy use (millions of liters petrol	Travel time (milions of bours)
Express ways	120	4.34 (27.6)	58.7 (43.2)	390 (4.3)	64 (6.2)	1848 (27.6)	265 (14.0)
Express ways	100	1.63 (10.3)	19.4 (14.2)	149 (1.6)	21 (2.0)	692 (10.3)	124 (6.6)
Highways	100	0.69 (4.3)	7.7 (5.7)	230 (2.5)	48 (4.7)	293 (4.3)	58 (3.1)
Highways	80	2.25 (14.3)	19.3 (14.2)	1792 (19.6)	318 (30.9)	955 (14.3)	282 (14.9)
Other roads outside urban areas	80	1.88 (11.9)	12.5 (9.2)	1834 (20.1)	282 (27.4)	798 (11.9)	289 (15.2)
Major urban roads	50	3.6 (22.7)	14.9 (11.0)	3731 (40.9)	233 (22.6)	1522 (22.7)	645 (34.0)
Other urban roads	50	1.38 (8.8)	3.5 (2.6)	965 (10.6)	61 (5.9)	587 (8.8)	225 (11.9)
Residential area (low speed)	30	0.02 (0.0)	0.02 (0.0)	36 (0.0)	2 (0.0)	9 (0.0)	7 (0.0)
All roads		15.76 (100.0)	136.0 (100.0)	9127 (100.0)	1029 (100.0)	6704 (100.0)	1895 (100.0)

[a] Based on an occupancy rate of 1.3 persons per car.

of the external effects, we find that NO_x is the most important issue on roads with maximum speeds of 100 km/h and above. Energy use and CO_2 assume the second position, and safety is relatively unimportant. For roads with a speed limit of 80 km/h safety is most important (especially when fatalities are used as an indicator). Energy and NO_x are less important. For major urban roads safety is most important (especially seriously injured persons); energy assumes the second position. From this table it becomes clear that the urgency of the various external effects is quite different from one road type to the other and that energy assumes an intermediate position in the sense that it is never the least or the most important issue.

The high safety level of the express ways is striking: the share of about 8% of the fatalities is very low compared with the share of about 36% of vehicle-km. This seems to contradict the increasing relationship between speed and fatalities. The explanation is of course that this increasing relationship does not necessarily hold when one compares different road types. Note for example that the concept of express ways precludes conflicts between cars and non-motorized traffic.

If one wants to consider the costs aspects of car use one can multiply the entries in Table 9.3 by the respective valuations shown in Table 9.1. The result for the medium valuation of the external effects (aggregated across all roads) shows that time losses are dominant among the costs distinguished here (a share of 70%), followed by petrol (11%, but remember that here only the costs of petrol without taxes are considered; the actual price paid by car drivers is four times as high). The shares of CO_2 emissions, NO_x, seriously injured persons and fatalities in total costs range between 4 and 6%. Of course the shares of these items vary considerably across road types. For example, on express roads the share of travel time is much lower than the aggregate 70% mentioned here.

9.5.2. *Direct effects*

On the basis of the above data and the models and parameters shortly sketched above we have been able to determine the impacts of changes of speed limits on the various environmental and safety effects for each road type. We distinguish four cases:[6]

- the present situation;
- the situation where drivers completely obey maximum speeds;
- the situation where maximum speeds would be set in such a way that a social cost benefit analysis would yield maximum net benefits according to a medium valuation of environmental and safety effects (see Table 9.1);
- the same result with a high valuation of environmental and safety effects (see Table 9.1).

The resulting average speeds for these cases are given in Table 9.4. The optimum

Table 9.4. Average speed (km/h) based on different speed regimes.

Road type	Present maximum speed (km/h)	Actual average speed (km/h)	Absolute obeyance of present speed limit	Optimal speeds given medium valuation of environmental effects	Optimal speeds given high valuation of environmental effects
Express ways	120	113	113	96	87
Express ways	100	105	96	96	87
Highways	100	92	92	86	73
Highways	80	80	73	73	73
Other roads outside urban areas	80	64	64	64	59
Major urban roads	50	38	35	35	35
Other urban roads	50	27	27	27	25
Residential area (low speed)	30	15	15	15	15
All roads		63	60	58	56

speeds given in this table are based on a summation of the speed–cost curves in Figures 9.6, where the curves are the result of speed–effect curves similar to those shown in Figures 9.3 and 9.4 multiplied by weights given in Table 9.1.

Strict obeyance of the present speed limits would lead to lower average speeds at three types of roads: the expressways with a present limit of 100 km/h, highways with a speed limit of 80 km/h, and major urban roads. As a consequence, the average speed of all trips would decrease from 63 to 60 km/h.

If we want to determine optimal speed limits based on a medium valuation of environmental effects, lower maximum speeds than the present maximum speeds would be found for 120 km/h express ways and for 100 km/h highways. On the other types of roads the maximum speeds would remain the same (100 km/h express roads) or might increase. With a high valuation of environmental effects, maximum speeds would have to be reduced for almost all types of roads (except major urban roads and roads in residential areas). In the last two columns of the table we find the results of a strict imposition of the current limits in conjunction with a strict imposition of lower limits if these would be favourable from a cost–benefit analytical point of view. The total reduction in average speed across all road types would in the most extreme case be about 13% (from the present level of 63 km/h to 56 km/h).

It should be noted, however, that due to data limitations, in these computations some elements have been ignored thus far which nevertheless would have an impact on the optimal level of speeds. First, noise effects have not been taken into account. These would be especially relevant in high density areas with intensive traffic flows; this will probably reinforce the case for speed reductions on major urban roads. In addition the list of pollutants taken into account is incomplete. Another point ignored thus far is that lower speeds will

Table 9.5. Effects of various speed regimes on environment, safety and time use in transport.

Type of effect	Present speeds	Strict obeyance of present speed limits	Optimal speed limits based on medium valuation of effects	Optimal speed limits based on high valuation of effects
Energy use	100	94	87	80
CO_2	100	94	87	80
NO_x	100	90	71	57
Number of seriously injured persons	100	90	89	83
Number of persons killed in transport	100	83	79	66
Travel time	100	105	108	113

probably lead to smaller levels of wear and tear, and hence to smaller expenditures for repair and maintenance. This would also reinforce the case for lower speed limits. The figures presented in Table 9.4 (and also of the following tables) are therefore based on a partial analysis of external effects.

The impacts of the speed regimes on the various types of environmental effects are described in Table 9.5. This clearly shows that the negative external effects of transport are relatively sensitive to changes in average speed. For example, a decrease of the average speed of 5% by strict obeyance of the present standard would lead to decreases in the various effects that range from 6% for energy and CO_2, a decrease of NO_x emissions of 10% and of traffic victims of 10–17%. This means that the elasticity of the various effects with respect to average speed is higher than 1. One should be aware, however, that the elasticity of the maximum speed with respect to the average speed is less than 1. In urban areas this elasticity will be particularly low because the average speed is strongly determined by patterns of stop and go in these areas. We find for example that a decrease of the maximum speed on 'other urban roads' from 50 to 40 km/h leads to a decrease of the average speed from 27 to 25 km/h. This implies an elasticity of about 0.4 of the average speed with respect to the maximum speed. For road types with a smaller number of interruptions this elasticity is higher. Note that these elasticities will also depend on the intensity of police surveillance.

We conclude that speed reduction policies when accompanied by appropriate surveillance are effective instruments to decrease negative externalities of transport.

9.5.3. *Indirect effects*

The above results are based on the assumption that speed changes do not lead to changes in travel behaviour. However, this is not a realistic assumption. By

Table 9.6. Impacts of various speed regimes on environment, safety and time use in transport; direct plus indirect effects.

Type of effect	Present speeds	Strict obeyance of present speed limits	Optimal speed limits based on medium valuation of effects	Optimal speed limits based on high valuation of effects
Number of kilometres driven	100	94	91	85
Energy use	100	89	79	68
CO_2	100	89	79	68
NO_x				
100	85	64	48	
Number of seriously injured persons	100	85	83	73
Number of persons killed in transport	100	79	75	60

making use of the approach outlined in Section 4 we have been able to approximate indirect effects on travel behaviour. The combined effects are given in Table 9.6. A remarkable feature of the outcomes is that the indirect effects are large. This is mainly caused by the high travel time elasticity of the LMS model of demand for road transport (−1.27) on which this outcome is based (Ministry of Transport, 1990). Effects of route choice may in principle lead to unfavourable unintended side effects. For example, speed reductions on express ways may lead to a shift to other types of roads so that traffic safety is negatively affected. In the present case it appears that such adverse effects are in most cases dominated by the effects of a decrease in the total number of kilometres driven. A point not taken into account in this table is that some of the car-km are substituted by public transport-km. Although public transport performs better than the private car in terms of emissions/passenger-km, it still certainly has unfavourable external effects which are ignored here. Another point is that when maximum speeds are reduced, this will lead to higher fuel efficiency of cars and to lower expenditures for maintenance and repair so that the costs per kilometre driven will decrease. This may have a countervailing indirect effect on the total number of kilometres driven. Also this effect has been ignored in this study.

Compared with the approach where only direct effects of speed changes are studied, the inclusion of indirect effects leads to higher net benefits. For example taking the present situation as a reference point the net benefits of the most extreme speed limits (sixth column in Table 9.4) increase from 1200 million Dutch guilders to 3200 million guilders. The difference with the case that only direct effects are taken into account can be explained by a larger decrease of environmental costs (fewer kilometres are travelled by car) and smaller travel time related welfare losses because some drivers prefer to respond to lower speeds by reducing kilometres travelled.

In this section we carried out an analysis of the broader consequences of the optimal speed limits found in Section 5.2. Inclusion of indirect effects into the analysis will probably lead to changes in the optimal speed limits for the various road types. Note that in Section 5.2 the assumption that the total number of kilometres driven per road type remains unaffected allows one to determine optimal speed levels for each road type independent from what takes place on other road types. This greatly facilitates the analysis. When indirect effects are also taken into account in the optimization procedure one would ideally optimize speed limits for various road types simultaneously. Such an optimization effort would make sense when a sufficiently detailed network model is available.

9.6. Conclusions

We have demonstrated that economic analysis provides a useful contribution to the discussion on speed limits. We have focused on two aspects: the need to address the whole range of roads (not only express ways), and the need to take into account behavioural changes of drivers as a result of changes in speeds.

Important limitations of the results presented here are that they are incomplete (for reasons of data availability not all relevant environmental effects have been taken into account) and that for the behavioural analysis one would like to use refined network models in order to study the route choice effects of speed regimes in an appropriate way. Nevertheless, although the exact levels of the figures may be subject to debate, the results make clear that changes in speeds have substantial effects on emissions, safety and welfare; hence speed limits are an important instrument for governments that want to reduce external effects of road transport.

In terms of the speed limits, our study indicates that the case for the imposition of lower limits from a cost–benefit point of view is strongest for the roads with the highest speeds (highways and expressways). This is in contrast with the opinions of drivers. Their support for speed reductions is rather low, but as far it exists it is strongest for the roads with lower speeds (Rienstra and Rietveld, 1996). An important reason for this discrepancy seems to be the visibility of the safety issue which is evident at roads with speeds up to 80 km/h and the invisibility of emissions and energy use which is especially important on roads with speeds of 80 km/h and above. A promising way to close this gap would be the introduction of special electronic equipment in cars which informs the driver about energy use and emissions ('econometer').

Speed policies should not only be considered from a national, but also from an international perspective. In Europe, where the share of cross border travelling is not a negligible part of total traffic, there is a case for an international coordination of speed policies. Especially when electronic means would be introduced to control speeds of cars, it is essential that standardization of technical specifications is carried out.

The optimal speed levels found in our study depend on a large number of parameters including:

- technological features of vehicles;
- behavioural responses of drivers;
- load factors of cars;
- value of travel time;
- valuation of external effects;
- design of roads;
- structure of road networks
- situational circumstances (weather).

Since these parameters are continuously changing during the course of time, it is advisable that speed regimes are evaluated from time to time.[7]

Another issue that deserves some attention is that in our approach we ignored heterogeneity of drivers. The optimal levels are based on the parameters of the average driver. However, the socially optimal speed of a driver may vary considerably among drivers, which is related among others to differences in the value of time of the driver and technological features of cars. Hence the question is how one can accommodate for this variety. One trend seems to be the further standardization of traffic on roads (car types become more similar), which would reduce the disadvantages of uniform maximum speeds. However, one can also foresee the emergence of the trend of customization in the road system. This would mean that a road operator would try to come to meet specific demands of drivers from the viewpoint of speeds. A possible way would be the introduction of pay-lanes on expressways where drivers have the option to choose between lanes with various speed regimes. Of course drivers who would use high speed lanes would have to pay special fares for this opportunity. To avoid excessive burdens on the environment, one might restrict the access to these roads to cars that satisfy certain technological standards. Some reflections on the issue of heterogeneity among drivers in the context of speed limits are given in Rietveld and Shefer (1996).

The approach followed in this paper can be extended and refined in several directions. In addition to the inclusion of the case of heterogeneity among drivers there is scope for further analyses based on among others more refined network models to deal with route choice aspects and models capable of dealing with other behavioural responses to speed regimes. Another important extension would be the inclusion of freight traffic into the analysis leading to heterogeneity of vehicles.

Notes

1. A reduction in engine size would be another way to reduce the maximum possible speed of cars. This would obviously have effects on roads where high speeds are possible (express ways), but also on roads where the maximum possible speed of cars would not be feasible. The reason is that energy use of such cars would be lower even when actual speeds at which they are driven do not change.
2. For simulation of speeds and energy use on express ways, use has been made of a more general simulation model called FOSIM (Vermijs, 1992)
3. Lave (1985), found that for US expressways speed variances have an impact on fatalities, whereas for average speed there is no statistically discernable relationship. His view has been challenged by Levy and Asch (1989), Fowles and Loeb (1989) and Snyder (1989); see Lave (1989) for a reply. It is clear that for other types of roads (with potential conflicts between cars coming from opposite directions, and cars meeting at crossings) speed per se certainly plays a role. Further, on expressways the severity of the conflict between cars will still depend on speed. Therefore we have chosen for a formulation for severe accidents and fatalities where both speed variance and average speed play a role (for a further discussion, see Shefer and Rietveld, 1997). Note that in our approach we assume different relationships between speed and fatalities for each type of road. These differences are due to the design of roads and the mixture of road users (for example, in the Netherlands cyclists are an important group of road users in urban areas). As a result on expressways relatively low accident rates are found. However, according to our model formulation, for each given type of road the accident and fatality rate increase with both speed and speed variance.
4. The Bleijenberg study, commissioned by the Dutch Ministry of Transport is the most recent Dutch study on this subject. In order to indicate the range of uncertainty on the valuations this study reports in addition to an average valuation also minimum and maximum values per type of external effect.
5. As long as business related trips are inelastic to speeds, as we assume at this stage of the analysis, this approach is sufficient to deal with the economic costs of speed reductions on business. This would no longer be true when these trips would be elastic, implying for example less frequent business visits, or a spatial restructuring of activity. This case of elastic demand is discussed in Section 5.3. Another issue would be the regulation of speeds for freight traffic. Our analytical approach can also be used for freight traffic, although the shapes of the curves involved would, of course, be quite different.
6. Note that we did not carry out an analysis based on a minimum valuation of external effects. If such an analysis would be carried out the result would be that it is optimal to increase maximum speeds on almost all types of roads. This is, however not a very meaningful result, because our analysis does not cover the whole range of external effects of road transport. Excluded are amongst others: small accidents, ozone, particulates and noise nuisance. It is no surprise to find that the combination of an incomplete coverage of external effects and a low valuation of the effects which are indeed valued leads to a general tendency to increase speeds. The effects ignored here seem to be of particular importance in urban areas. Thus, their inclusion would especially reinforce the case of speed reductions in urban areas.
7. From this perspective the introduction of a system with flexible speed limiting equiment would be quite attractive. It would not only help to solve the surveillance problem, but also allow the imposition of situation dependent speed limits (for example: low speeds in the case of fog, or near locations where road maintenance activities take place).

References

Bleijenberg, A.N., W.J. van den Berg and G. de Wit, 1994, Maatschappelijke kosten van het verkeer, Den Haag: Ministry of Transport.

ECMT (European Conference of Ministers of Transport), 1978, Costs and benefits of general speed limits, Paris: ECMT.
Fowles, R. and P.D. Loeb, 1989, Speeding coordination and the 55-mph limit: comment, American Economic Review, 79, 916–921.
HCG (Hague Consulting Group), 1990, The Netherlands' value of time study: final report, Den Haag: HCG.
Jorgenson, F. and J. Polak, 1993, The effect of personal characteristics of drivers' speed selection; an economic approach, Journal of Transport Economics and Policy, 27, 237–252.
Kageson, P., 1993, Getting the prices right, Brussels/Stockholm European Federation for Transport and Environment.
Klein, J.A.P., 1993, Luchtverontreiniging, emissies door wegverkeer, Statistische onderzoekingen M45, Den Haag: Central Bureau of Statistics.
Koornstra, M.J., 1993, Road Safety Effects of Changes in Speeds, Den Haag: SWOV.
Lave, C.A., 1985, Speeding coordination and the 55-mph limit, American Economic Review, 75, 1159–1164.
Lave, C.A., 1989, Speeding coordination and the 55-mph limit: reply, American Economic Review, 79, 926–931.
Levy, D.T. and P. Asch, 1989, Speeding coordination and the 55-mph limit: comment, American Economic Review, 79, 913–915.
Ministry of Transport, 1990, Het Landelijk Modelsysteem, Den Haag: Ministry of Transport.
Pearce, D. and A. Markandya, 1989, Environmental policy benefits: monetary valuation, Paris: OECD.
Peeters, P.M. and Y. van Asseldonk, 1996, Mag het ietsje minder snel? Een onderzoek naar de maatschappelijke economische kosten en baten van verlaging van snelheden van personenauto's, Den Haag: IVVS.
Quinet, E., 1990, The social costs of land transport, Paris: OECD.
Rienstra, S. and P. Rietveld, 1996, Speed behaviour of car drivers; a statistical ananlysis of acceptance of changes in speed policies in the Netherlands, Transportation Research D (Transport and the Environment), 1, 1996, pp. 97–110.
Rietveld, P. and D. Shefer, 1996, A note on speed limits as a second best instrument to correct for road transport externalities, Amsterdam: Tinbergen Institute.
Rouwendal, J., 1996, An economic analysis of fuel use per kilometre by private cars, Journal of Transport Economics and Policy, 30, 3–14.
Shefer, D. and P. Rietveld, 1997, Congestion and safety on highways: towards an analytical model, Urban Studies, 34, 679–692.
Snyder, D., 1989, Speeding coordination and the 55-mph limit: comment, American Economic Review, 79, 922–925.
Technical University Delft (TUD), 1995, Berekening van maatschappelijke kosten en baten van snelheidsverlaging voor het personenautoverkeer, Delft: TUD.
Verhoef, E., 1994, External effects and social costs of road transport, Transportation Research A, 28, 273–287.
Vermijs, G.G.M.M., 1992, Het microsimulatiemodel FOSIM, Vakgroep Infrastructuur, Delft: Technical University.

List of Contributors

Anna Alberini, Department of Economics, University of Colorado, Boulder, CO, USA.

John Bates, John Bates Services, Oxford, UK.

Bruno De Borger, UFSIA-SESO, University of Antwerp, Belgium.

Winston Harrington, Resources for the Future, Washington, DC, USA.

Glenn W. Harrison, Department of Economics, University of South Carolina, Columbia, SC, USA.

Bengt Kriström, Department of Forest Economics, Swedish University of Agricultural Science, Umeå, Sweden.

Virginia McConnell, University of Maryland at Baltimore and Resources for the Future, Washington, DC, USA.

Paul Peeters, Werkgroep 2000, Amersfoort, The Netherlands.

John Peirson, Centre for European, Regional and Transport Economics, The University of Kent at Canterbury, UK.

Giancarlo Pireddu, Scuola Superiore Enrico Mattei, Milan, Italy.

Piet Rietveld, Free University, Amsterdam, The Netherlands.

Roberto Roson, Department of Economics, Ca' Foscari University, Venice, Italy.

Theo Schoemaker, Technische Universiteit, Delft, The Netherlands.

Kenneth A. Small, Department of Economics, University of California, Irvine, CA, USA.

Didier Swysen, UFSIA-SESO, University of Antwerp, Belgium.

Arjan van Binsbergen, Technische Universiteit, Delft, The Netherlands.

Roger Vickerman, Centre for European Regional and Transport Economics, The University of Kent at Canterbury, UK.

R. Roson and K.A. Small (eds.), Environment and Transport in Economic Modelling, 226.

Subject Index

R. Roson and K.A. Small (eds.), Environment and Transport in Economic Modelling, 227–234.

FONDAZIONE ENI ENRICO MATTEI (FEEM) SERIES ON ECONOMICS, ENERGY AND ENVIRONMENT

This series serves as an outlet for the main results of FEEM's research programmes in the areas of economics, energy and environment.

---

1. C. Carraro and D. Siniscalco (eds.), *The European Carbon Tax: An Economic Assessment.* 1993 ISBN 0-7923-2520-6
2. C. Carraro (ed.), *Trade, Innovation, Environment.* 1994 ISBN 0-7923-3033-1
3. C. Dosi and T. Tomasi (eds.), *Nonpoint Source Pollution Regulation: Issues and Analysis.* 1994 ISBN 0-7923-3121-4
4. C. Carraro, Y. Katsoulacos and A. Xepapadeas (eds.), *Environmental Policy and Market Structure.* 1996 ISBN 0-7923-3656-9
5. C. Carraro and A. Haurie (eds.), *Operations Research and Environmental Management.* 1996 ISBN 0-7923-3767-7
6. I. Musu and D. Siniscalco (eds.), *National Accounts and the Environment.* 1996 ISBN 0-7923-3741-7
7. C. Carraro and D. Siniscalco (eds.), *Environmental Fiscal Reform and Unemployment.* 1996 ISBN 0-7923-3750-6
8. A. Beltratti: *Models of Economic Growth with Environmental Assets.* 1996 ISBN 0-7923-4032-9
9. G. Chichilnisky, G. Heal and A. Vercelli (eds.), *Sustainability: Dynamics and Uncertainty.* 1998 ISBN 0-7923-4698-X
10. R. Roson and K.A. Small (eds.), *Environment and Transport in Economic Modelling.* 1998 ISBN 0-7923-4913-X

KLUWER ACADEMIC PUBLISHERS – DORDRECHT / BOSTON / LONDON